Manchester were the city infirmary (founded in 1752) and the Cross Street Unitarian Chapel. The doctors and the Unitarians were the intellectual elite of the rapidly growing town, and they joined forces in 1781 to create an institution appropriate to their needs, the Manchester Literary and Philosophical Society. The Literary and Philosophical Society was consciously designed to give Manchester a cultural focus, to raise the aspirations of the leading citizens to a loftier level, to divert them from the mere accumulation of wealth to the pursuit of higher learning. From the very beginning the society was enormously successful. It attracted and supported first-rate scientists such as Priestley, Dalton and Joule, it published a journal, it built a library and a College of Arts and Sciences and a Mechanics' Institution and an Academy, it gave birth to Owens College which ultimately grew into the University of Manchester, and it had a big share in the founding of the British Association for the Advancement of Science. During the first seventy years of its existence, from 1781 to 1851, the society had altogether 588 members, and of these no less than 31, a little over 5 percent, became sufficiently distinguished as scientists on the national scene to be elected Fellows of the Royal Society of London. An extraordinary achievement for a bunch of amateurs in a raw provincial town.

The historical question which needs to be answered is why a group of doctors and Unitarian Chapel-goers, having decided to give their city a new cultural identity, should have found it in the study of physics and chemistry. The name of the Literary and Philosophical Society shows that their original objective was general culture and not specialized science. Thomas Henry, who was both a physician and a Unitarian and one of the founding members of the society, expressed their purpose explicitly: "A taste for polite literature, and the works of nature and art, is essentially necessary to form the gentleman." In other words, the founding fathers wanted to prove that it was possible to live in Manchester and still be a gentleman. How did it happen that the search for genteel status led them so rapidly and decisively into science?

Arnold Thackray explains their concentration upon science as a result of two main factors. First, the Unitarians were legally barred from the academic establishment of Oxford and Cambridge, and the atmosphere of Manchester was saturated with contempt for the ancient universities. So the organizers of the Literary and Philosophical Society were anti-academic, having no use for the smattering of Latin and Greek which the universities in those days called a classical education. Second, the inhabitants of Manchester were unrepresented in Parliament and therefore tended to radicalism in politics, especially in the formative years of the society before the French Revolution made radicalism unpopular. Radical politics included a belief in public education, and science served better than Latin and Greek as a vehicle for educating the masses. The chemist Priestley, hero of the radicals, expressed their view of science as an agent of social reform: "The English Hierarchy, if there be anything unsound in its constitution, has reason to tremble even at an air pump or an electrical machine."

So the anti-academic, anti-establishment brashness of Manchester made a fertile ground for the growth of science. And the science which grew in that northern soil had a style different from the science of Athens, just as two hundred years later the music of the Beatles growing up in nearby Liverpool had a style different from the music of Mozart. The science of Athens emphasizes ideas and theories; it tries to find unifying concepts which tie the universe together. The science of Manchester emphasizes facts and things; it tries to explore and extend our knowledge of nature's diversity. Of course the tradition of unifying science did not end with Athens, and the tradition of diversifying science did not begin with Manchester. Historians of science are accustomed to call these two traditions in science Cartesian and Baconian, since Descartes was the great unifier and Bacon the great diversifier at the birth of modern science in the seventeenth century. The unifying and diversifying traditions have always remained alive in science to a greater or lesser extent. But the human exploit which Disraeli discerned in Manchester included an important rebirth of the diversifying tradition in science. Manchester brought science out of the academies and gave it to the people. Manchester insolently repudiated the ancient prohibition, "Let nobody ignorant of geometry enter here," which Plato is said to have inscribed over the door of his academy in Athens.

We now come to the modern era. The years 1907 to 1919 were the high noon of physics in Manchester. During those years Ernest Rutherford was professor in Manchester, at the peak of his youthful vigor, creating the science that later came to be called nuclear physics. In Manchester he discovered the atomic nucleus and observed the first nuclear reactions. He was a scientist in the tradition of the Manchester Literary and Philosophical Society, self-confident and largely self-taught, disrespectful of academic learning, more interested in facts than in theories. Concerning theoretical physicists he made the famous statement, "They play games with their symbols, but we turn out the real facts of Nature." In 1913 the great theoretician Niels Bohr sent to Rutherford in Manchester the manuscript of his epoch-making paper on the quantum theory of the atom. Rutherford understood the importance of Bohr's work and volunteered to send the paper to the *Philosophical Magazine* for publication. But at the end of Rutherford's letter to Bohr there is a brief postscript: "I suppose you have no objection to my using my judgement to cut out any matter I may consider unnecessary in your paper? Please reply." Rutherford stood in awe of nobody. Not even of Einstein.

The astronomer Subrahmanyan Chandrasekhar tells a story of a conversation which he heard in 1933 between Rutherford and Arthur Eddington. Eddington was the astronomer who obtained the crucial evidence of the correctness of Einstein's general relativity by observing the gravitational deflection of light-rays by the Sun at the total eclipse of 1919. A friend said to Rutherford in the presence of Eddington and Chandrasekhar, "I do not see why Einstein is accorded a greater public acclaim than you. After all, you invented the nuclear model of the atom, and that model provides the basis for all of physical science today. . . ." Rutherford, in response, turned to Eddington and said, *"You* are responsible for Einstein's fame." Rutherford was not belittling Einstein's achievement, but he went on to say that the typhoon of publicity surrounding Einstein was a result of the dramatic circumstances of the eclipse expeditions and of the dramatic manner in which Eddington's results were announced in London. Clearly, Rutherford felt that Einstein's fame was somewhat disproportionate, and he was not afraid to say so.

The point of telling this story is not to pull down Einstein from his position as the greatest scientist of our century. Chandrasekhar ends his account by remarking that the true and lasting preeminence of Einstein lies in the incredibly rich content of the general theory of relativity itself. Chandrasekhar agrees with Einstein's own triumphant statement in his first presentation of the theory to the Berlin Academy, "Scarcely anyone who has fully understood this theory can escape from its magic." The point of the story is that Rutherford did not understand general relativity and was therefore immune to its magic. Rutherford's science and Einstein's science were so different in style that no real understanding between them was possible. The gulf between them was deeper than the normal gulf between experimenters and theorists. The problem was not merely that Einstein did not care about alpha-particles and Rutherford did not care about curvature-tensors. The problem was that they had fundamentally different concepts of the nature and purpose of science. Einstein said: "The creative principle resides in mathematics. In a certain sense, therefore, I hold it true that pure thought can grasp reality, as the ancients dreamed." Rutherford said: "Continental people do not seem to be in the least interested to form a physical idea as a basis of theory. They are quite content to explain everything on a certain assumption and do not bother their heads about the real cause of a thing. I must say that the English point of view is much more physical and much to be preferred." And on another occasion when Eddington remarked at the dinner table that electrons were very useful conceptions but might not have any real existence, Rutherford replied, "Not exist, not exist,—why I can see the little beggars there in front of me as plainly as I can see that spoon."

So there are the two styles of science typified by Athens and Manchester, Einstein and Rutherford, abstract and concrete, unifying and diversifying. The two styles are not in conflict with one another. They are complementary, giving us two views of the universe which are both valid but cannot both be seen simultaneously. The word "complementary" has here the technical meaning that it has in Niels Bohr's interpretation of quantum mechanics. According to Bohr, an electron cannot be pictured as a simple material object but must be described by two complementary pictures emphasizing its particle behavior and its wave behavior separately. Einstein and Rutherford gave us complementary views of science, and each was too single-mindedly attached to his own view to understand the other. Both of them, for opposite reasons, rejected the compromise which Bohr's notion of complementarity offered them. For Einstein, the electron must ultimately be understood as a clumping of waves in a non-linear field theory. For Rutherford, the electron remained a particle, a little beggar that he could see in front of him as plainly as a spoon.

The discoveries of recent decades in particle physics have led us to place great emphasis on the concept of broken symmetry. The development of the universe from its earliest beginnings is regarded as a succession of symmetry-breakings. As it emerges from the moment of creation in the Big Bang, the universe is completely symmetrical and featureless. As it cools to lower and lower temperatures, it breaks one symmetry after another, allowing more and more diversity of structure to come into existence. The phenomenon of life also fits naturally into this picture. Life too is a symmetry-breaking. In the beginning a homogeneous ocean somehow differentiated itself into cells and animalcules, predators and prey. Later on, a homogeneous population of apes differentiated itself into languages and cultures, arts and sciences and religions. Every time a symmetry is broken, new levels of diversity and creativity become possible. It may be that the nature of our universe and the nature of life are such that this process of diversification will have no end.

If this view of the universe as a steady progression of symmetry-breakings is valid, then Athens and Manchester fit in a natural way into the picture. The science of Athens, the science of Einstein, tries to find the underlying unifying principles of the universe by looking for hidden symmetries. Einstein's general relativity showed for the first time the enormous power of mathematical symmetry as a tool of discovery. Now we have reason to believe that the symmetry of the universe becomes explicit and the laws of its behavior become unified if we go back far enough into the past. Particle physics is at the moment at the threshold of a big new step in this direction with the construction of Grand Unified models of the strong and weak interactions. The details of the Grand Unified models are worked out by studying the dynamics and composition of the universe as it is presumed to have existed for an unimaginably small fraction of a second after its beginning. Whether or not the particular models now proposed turn out to be correct, there is no doubt that the concept of unifying physics by going back to a simpler and more symmetric past is a fruitful one. The science of Athens, what Einstein called the ancient dream that pure thought can grasp reality, is then nothing else than the exploration of our remotest past, "À la recherche du temps perdu" on a bolder scale than Proust ever imagined.

In a similar fashion, the science of Manchester and of Rutherford, the science of the diversifiers, is an exploration of the universe oriented toward the future. The further we go into the future, the more diversity of natural structures we shall discover, and the more diversity of technological artifice we shall create. It is then easy to understand why we have two kinds of scientists, the unifiers looking inward and backward into the past, the diversifiers looking outward and forward into the future. Unifiers are people whose driving passion is to find general principles which will explain everything. They are happy if they can leave the universe looking a little simpler than they found it. Diversifiers are people whose passion is to explore details. They are in love with the heterogeneity of nature and they agree with the saying, "Le bon Dieu aime les détails." They are happy if they leave the universe a little more complicated than they found it.

Now it is generally true that the very greatest scientists in each discipline are unifiers. This is especially true in physics. Newton and Einstein were supreme as unifiers. The great triumphs of physics have been triumphs of unification. We almost take it for granted that the road of progress in physics will be a wider and wider unification bringing more and more phenomena within the scope of a few fundamental principles. Einstein was so confident of the correctness of this road of unification that at the end of his life he took almost no interest in the experimental discoveries which were then beginning to make the world of physics more complicated. It is difficult to find among physicists any serious voices in opposition to unification. One such voice is that of Emil Wiechert: "If we start from our human scale of existence and explore the content of the universe further and further, we finally arrive, both in the large and in the small, at misty distances where first our senses and then even our concepts fail us."

This remark of Wiechert's shows him to have been extraordinarily far-sighted. At the time when he was speaking, the leading theoretical physicists of Germany were still engaged in bitter arguments over the question of the real existence of atoms. Rutherford was then a young student newly arrived in England from New Zealand, only just started on the road that would lead him to the discovery of alpha-particles and atomic nuclei. More than half a century was to pass by before the flowering of experimental particle physics, which would reveal a whole new world of strange objects and strange interactions hidden within structures that are themselves far smaller than atoms. Wiechert's words were ignored. His vision was too large for the time he lived in.

In biology the roles are reversed. A very few of the greatest biologists are unifiers. Darwin was a unifier, consciously seeing himself as achieving for biology the unification which Newton had achieved for physics. Darwin succeeded in encompassing the entire organic world within his theory of evolution. But the organic world remains fundamentally diverse. Diversity is the essence of life, and the essential achievement of Darwin's theory was to give intellectual coherence to that diversity. The working lives of ninety-nine out of a hundred biologists are spent in exploring the details of life's diversity, disentangling the complex behavior patterns of particular species or the marvelously intricate architecture of particular biochemical pathways. Biology is the natural domain of diversifiers as physics is the domain of unifiers. Unifiers like Darwin are as rare in biology as diversifiers like Wiechert are rare in physics. Darwin had no peer and no successor.

Or perhaps I should say, Darwin has only one successor and his name is Francis Crick. In saying this, I am not expressing a judgment of the greatness of Crick as compared with other contemporary biologists, and still less am I expressing a judgment of the importance of molecular biology as compared with botany and zoology. I am merely saying, Crick is a unifier of biology in the style and tradition of Darwin.

Ninety-eight years after Darwin published his *Origin of Species,* Crick propounded and named the Central Dogma of molecular biology:

"The Central Dogma. This states that once information has passed into protein it cannot get out again. In more detail, the transfer of information from nucleic acid to nucleic acid, or from nucleic acid to protein may be possible, but transfer from protein to protein, or from protein to nucleic acid is impossible."

The propounding of dogmas is an unusual activity for a biologist. Even Crick does not spend much of his time propounding dogmas. He spends most of his time studying the details of particular structures, just as Darwin spent much of his time studying details of the taxonomy of barnacles. But at heart Crick is a unifier, and in the end he will be remembered as the man who stated in simple words the unifying principle of biology for the twentieth century, as Darwin did for the nineteenth.

Let me now pull together the threads of my argument. I am saying that every science needs for its healthy growth a creative balance between unifiers and diversifiers. In the physics of the last hundred years, the unifiers have had things too much their own way. Diversifiers in physics, such as Wiechert in the 1890s and John Wheeler in our own time, have tended to be pushed out of the mainstream. John Wheeler is a Princeton professor whose style is so unorthodox that he tends to publish his ideas in books rather than in conventional journal articles. He holds uncompromisingly to his diversifier's view of the physical universe. Here is the theme song of Wheeler's recent book, *Frontiers of Time:*

"Individual events. Events beyond law. Events so numerous and so uncoordinated that, flaunting their freedom from formula, they yet fabricate firm form."

It sounds like *Beowulf,* but it is authentic Wheeler. This vision of nature is regarded by orthodox physicists as belonging to poetry rather than to science. Wheeler's colleagues love him more than they listen to him. The physics of the unifiers has no room for his subversive thoughts.

In biology there has been a healthier balance. The mainstream of biology is the domain of the diversifiers, the domain of events numerous and uncoordinated, flaunting their freedom from formula. But when a unifier like Darwin or Crick arrives on the scene, he is not ignored. He is even, after a while, honored and rewarded. And his ideas flow into the mainstream. I am suggesting that there may come a time when physics will be willing to learn from biology as biology has been willing to learn from physics, a time when physics will accept the endless diversity of nature as one of its central themes, just as biology has accepted the unity of the genetic coding apparatus as one of its central dogmas.

The history of science is full of dichotomies. Athens and Manchester, past and future, unity and diversity, the universe of the cosmologist and the universe of the biologist. Cosmology is the study of the abstract structure of the universe in the large, biology is the study of its most intricate details. At various times in the historical development of science, one side or the other of these dichotomies has been overemphasized. Sometimes unity and abstract structure are overemphasized. Then the universe is seen as the solution of a finite set of equations, the equations of mathematical cosmology and the equations of superstrings, and once we have these equations right the remaining tasks of science are regarded as little more than butterfly-collecting. Sometimes diversity and richness of detail are overemphasized. Then the universe is seen as infinite in all directions, but without any backbone of mathematical structure to give it coherence.

There was once a time when the ideals of unity and diversity in science were briefly held in balance. This was in the seventeenth century, when modern science was in its first flowering and both Descartes and Bacon were honored. There was then no clear separation between the sciences of cosmology and biology. Even the most austere and respectable physicists conceived of the heavenly universe as filled with living creatures. Christiaan Huygens, originator of the wave theory of light and a physicist of impeccable credentials, wrote a book on cosmology with the title *Cosmotheoros.* "A man that is of Copernicus's opinion," he wrote, "that this Earth of ours is a Planet, carry'd round and enlighten'd by the Sun, like the rest of the Planets, cannot but sometimes think that it's not improbable that the rest of the Planets have their Dress and Furniture, and perhaps their Inhabitants too as well as this Earth of ours." His great contemporary Isaac Newton, the supreme intellect of the Age of Enlightenment, carried such thoughts even further:

> As all regions below are replenished with living creatures, (not only the Earth with Beasts, and Sea with Fishes and the air with Fowls and Insects, but also standing waters, vineger, the bodies and blood of Animals and other juices

with innumerable living creatures too small to be seen without the help of magnifying glasses) so may the heavens above be replenished with beings whose nature we do not understand. He that shall well consider the strange and wonderful nature of life and the frame of Animals, will think nothing beyond the possibility of nature, nothing too hard for the omnipotent power of God. And as the Planets remain in their orbs, so may any other bodies subsist at any distance from the earth, and much more may beings, who have a sufficient power of self motion, move whether they will, place themselves where they will, and continue in any regions of the heavens whatever, there to enjoy the society of one another, and by their messengers or Angels to rule the earth and convers with the remotest regions. Thus may the whole heavens or any part thereof whatever be the habitation of the Blessed, and at the same time the earth be subject to their dominion. And to have thus the liberty and dominion of the whole heavens and the choice of the happiest places for abode seems a greater happiness then to be confined to any one place whatever.

I quote this passage from a recent book, *The Religion of Isaac Newton*, by Frank Manuel. Huygens and Newton belonged to the last generation of cosmologists who felt free to people the universe with creatures in this fashion. The style and temper of science were already changing in a direction which would make such flights of fancy unacceptable. Neither Huygens nor Newton had the courage to expose their speculations to the ridicule of the public. Huygens arranged for his *Cosmotheoros* to be published after he was safely dead. Newton was even more timid and never published his private thoughts at all. But he was careful to preserve his manuscripts, and they can now after three hundred years be found, still unpublished, in the library of the Hebrew University in Jerusalem. As Frank Manuel remarks: "This extensive text proves beyond question that Newton's world-view in the decade when the *Principia* was composed admitted of a far greater diversity of beings than those recognized by positivist physical scientists and nineteenth-century Unitarians." But Newton's manuscripts stayed hidden from his contemporaries in a big black box, and the eighteenth century knew nothing of them. The eighteenth century dawned bleakly under a heaven grown empty and dead. Cosmology, ever since that time, has concerned itself only with an empty and dead universe. When Newton decided to suppress his youthful visions of the cosmos, he was only doing what every good scientist is supposed to do, abandoning without mercy a beautiful theory which turned out to be unsupported by experimental facts.

So we have been left since Newton's time with a cosmology in which living creatures play no part. Only a few heretics like Emil Wiechert and John Wheeler dare to express the view that the structure of the universe may not be unambiguously reducible to a problem in physics. Only a few romantics like me continue to hope that one day the links between biology and cosmology may be restored.

What can we do now to start building bridges between biology and cosmology? There are at least two things we can do. The first thing is to look very hard at the universe and search for evidence of life and intelligence in remote places. If we are lucky, we may find that Newton gave up too easily his universe of celestial beings, that the cosmos is not really so dead and empty as it looks. The search for evidence of extraterrestrial intelligence is a continuing enterprise in which many respectable astronomers are intermittently engaged. But the job of searching the universe for traces of life is a job for observers, not for theorists. There is little that theorists can do to help, except to serve as fund raisers and cheerleaders.

There is a second way of building bridges between biology and cosmology, a way open to theorists rather than to observers. This second way is open, whether or not the universe turns out to be peopled with celestial friends and colleagues. The second way is to build general theories of the potentialities of life in the universe. Francis Crick has been active here too. A few years ago I received a letter from him saying, "I am still interested in the idea of Directed Panspermia. Our slogan was 'Bugs can go further.' " By this he means that there is every reason to expect spores to play an essential role in the propagation of life in the universe just as they do in the propagation of life on Earth. Spores are the natural way to package biological and genetic information for rapid transit over interstellar distances. Panspermia is an old theory, originally proposed by the chemist Svante Arrhenius, a contemporary of Emil Wiechert. Arrhenius imagined the whole universe filled with the spores of life. Directed Panspermia is panspermia plus intelligence, the universe filled with spores deliberately aimed toward habitats favorable to life's spread and survival.

Directed panspermia is only a hypothesis on the wilder fringe of speculation, not quite science and not quite science fiction. It belongs with Newton's celestial zoo in the borderland where science and mythology meet. Future-oriented science easily runs astray into undisciplined speculation and outright fiction. Newton's timid soul, afraid to let his imagination run free, retreated to the safe ground of conventional science and conventional theology. Fortunately our age is more tolerant than Newton's of scientific heretics. The prospects are bright for a future-oriented science, joining together in a disciplined fashion the resources of biology and cosmology. When this new science has grown mature enough to differentiate itself clearly from the surrounding farrago of myth and fiction, it might call itself "cosmic ecology," the science of life in interaction with the cosmos as a whole. Cosmic ecology would look to the future rather than to the past for its subject matter, and would admit life and intelligence on an equal footing with general relativity as factors influencing the evolution of the universe. I will come back in Chapter 6 to the subject of cosmic ecology and the long-range future of life.

One of the people who built bridges in this century between cosmology and biology was Michael Polanyi. Polanyi was in Aberdeen as Gifford Lecturer in 1951 and 1952. His lectures were published in a book with the title *Personal Knowl-*

edge. Polanyi was a chemist by profession and a philosopher by temperament. He had a diversifier's view of science. He moved easily between chemistry and philosophy, between the industrial world and the academic world. As a young man in Karlsruhe, he supported himself by solving problems for the chemical industry. It is no accident that he fitted the style and spirit of Manchester when he came there as a refugee from Hitler at the age of forty-two. It is no accident that he found his spiritual home in Manchester rather than in the purer academic atmosphere of Oxford or Cambridge. The main theme of my Gifford Lectures is summed up in one sentence of his: "This universe is still dead, but it already has the capacity of coming to life."

Among contemporary physicists, John Wheeler is unique in taking seriously the possibility that the laws of physics may be contingent upon the presence of life in the universe. Wheeler says, "It is preposterous to think of the laws of physics as installed by a Swiss watchmaker to endure from everlasting to everlasting when we know that the universe began with a big bang. The laws must have come into being. Therefore they could not have been always a hundred percent accurate. That means that they are derivative, not primary. . . . Of all strange features of the universe, none are stranger than these: time is transcended, laws are mutable, and observer-participancy matters." According to Wheeler, the laws of physics evolve progressively in such a way as to make the universe observable.

Fifty years ago, Kurt Gödel, who afterwards became one of Einstein's closest friends, proved that the world of pure mathematics is inexhaustible. No finite set of axioms and rules of inference can ever encompass the whole of mathematics. Given any finite set of axioms, we can find meaningful mathematical questions which the axioms leave unanswered. This discovery of Gödel came at first as an unwelcome shock to many mathematicians. It destroyed once and for all the hope that they could solve the problem of deciding by a systematic procedure the truth or falsehood of any mathematical statement. After the initial shock was over, the mathematicians realized that Gödel's theorem, in denying them the possibility of a universal algorithm to settle all questions, gave them instead a guarantee that mathematics can never die. No matter how far mathematics progresses and no matter how many problems are solved, there will always be, thanks to Gödel, fresh questions to ask and fresh ideas to discover.

It is my hope that we may be able to prove the world of physics as inexhaustible as the world of mathematics. Some of our colleagues in particle physics think that they are coming close to a complete understanding of the basic laws of nature. They have indeed made wonderful progress in the last ten years. But I hope that the notion of a final statement of the laws of physics will prove as illusory as the notion of a formal decision process for all of mathematics. If it should turn out that the whole of physical reality can be described by a finite set of equations, I would be disappointed. I would feel that the Creator had been uncharacteristically lacking in imagination. I would have to say, as Einstein once said in a similar context, "Da könnt' mir halt der liebe Gott leid tun" ("Then I would have been sorry for the dear Lord").

Fortunately, the recent successes of particle physics and of cosmology do not exclude the possibility that the world of physics is truly inexhaustible, that Michael Polanyi was right when he said: "This universe is still dead, but it already has the capacity of coming to life," that John Wheeler is right when he says: "The universe is a self-excited circuit," that Emil Wiechert was right when he said: "The universe is infinite in all directions."

4

HOW DID LIFE BEGIN?

I am concerned with the origin of life as a scientific problem, not as a philosophical or theological problem. The problem is to find out what happened. To understand what happened, it is not necessary to agree upon a verbal definition of life. So far as science is concerned, the problem is to find a plausible sequence of events by which a barren and lifeless planet becomes transformed into the planet we see around us today. The question, at which point along that sequence of events we draw the line between non-living and living, is not a scientific question. The line between living and non-living at the beginning of evolution is arbitrary, just as the line between human and non-human primates at the end of evolution is arbitrary. In both cases, we are free to draw the line according to our taste. In both cases, the task of science is not to define the exact position of the line but to understand how it came to be crossed.

A hundred years ago, Charles Darwin could write books discussing the central problems of biology in language which was scientifically precise and still accessible to the general public. In those days the subject matter of biology was plants and animals. The language of Darwin was intelligible to experts and non-experts alike. One did not need a degree in botany to understand the difference between a fern and a flower. Darwin could assume that his readers were familiar with the world of nature as he described it.

Today, unfortunately, the language of Darwin is no longer adequate to describe the main themes of science. The language of physics has moved steadily further into the domain of abstract mathematics. Instead of describing nature with mechanical models, physicists now describe it with infinite-dimensional spaces and other even more esoteric mathematical concepts. And the language of biology has become heavily encrusted with the jargon of chemistry. During the last fifty years, biologists have made enormous progress in understanding the basic processes of life. The basic processes are chemical reactions. The price of progress is a language in which "amino acid," "nucleotide," "protein," "nucleic acid" are the common nouns, "synthesize," "catalyze," "polymerize" the common verbs. The chemistry of living cells is the essence of life. The beauty of their variegated architecture is inescapably tied to the ugliness of chemical terminology. The further we penetrate into the mysteries of their behavior, the longer and the uglier our words become.

Anyone who tries to explain modern science to the general public faces a choice, either to write an elementary textbook with technical terminology carefully defined, or to take refuge in metaphor. I have chosen the latter course. In Chapter 1 I used the metaphor of the rain forest to describe the proliferation of particles and of theoretical models in modern physics. The metaphor is impressionistic, but it is not altogether misleading. It may be more illuminating than an elementary description of the objects, quarks and leptons and hadrons and so forth, which the experts are exploring. If the reader would prefer an elementary textbook, there are plenty available, both in particle physics and in molecular biology. I do not need to write another. I shall use metaphor again to describe the chemical architecture of life. In this way I avoid the multitude of technical definitions which are needed if one names each brick in the structure by its proper name.

The metaphor of the computer represents in some crude fashion the chemistry of life. Nowadays one may assume that the average citizen of an industrialized country is at least as familiar with computers as with rain forests. The idea of using the computer as a metaphor is a natural one. A computer is a device for handling information according to a program which it is able to remember and execute. A living cell, to remain in control of its vital functions in a variable environment, must also possess a program of chemical reactions which it is able to remember and execute. The control system of the cell must be capable of storing and handling the large amount of information that is needed to keep a large number of complicated chemical reactions in balance.

Everyone who works with computers knows that two essential components are needed to make them useful. The components are hardware and software. Hardware is the computer itself, the box full of electronic circuits which carry out logical or mathematical operations. Software is the floppy disk on which instructions and information are written. You feed the software into the hardware in order to tell the hardware what to do. Likewise, every living cell has two dominant components, two kinds of large chemical molecules called proteins and nucleic acids. The discoveries of the last fifty years have revealed that proteins behave like hardware, acting as catalysts to make other chemicals react in highly specific ways. And nucleic acids act like software, organizing the proteins and telling them what to do. The decisive event which started the modern era in biology was an experiment done by Oswald Avery and his colleagues in New York in 1944. Avery took bacteria of one variety and fed them nucleic acid from another variety. The bacteria were transformed into the variety from which the nucleic acid came. If computers and floppy disks had existed in 1944, Avery might have said that the change of nucleic acid was affecting his bacteria just as a change of disk affects a computer.

The computer metaphor has been enormously fruitful as an aid to the understanding of living processes. Like the metaphor of the rain forest, it is not exact. It is only a metaphor. One should not push metaphors too far. One should not think that a molecule of nucleic acid is really an inert object like a floppy disk. In reality, a molecule of nucleic acid can do many other things besides serving as a passive container of information. Recent experiments have shown that under some conditions nucleic acids can act as catalysts just as proteins do. The separation between hardware and software in a living cell is not absolute. There is no law of nature which forbids proteins to carry information and forbids nucleic acids to process information. Nevertheless, the metaphor is a good one to describe the overall organization of a cell. Most of the time, proteins are busy chopping and splicing while nucleic acids are quietly telling them what to do.

The computer metaphor can be extended a little further. Living cells have two primary functions which are given the names "metabolism" and "replication." Metabolism means eating and digesting and excreting. Metabolism maintains the integrity of the cell by a continual reshuffling of chemical components, converting raw materials from outside the cell into the substances required for its continued existence. Replication means making exact copies of molecules so that a cell can reproduce itself. Replication allows the copying of hereditary information so that the characteristics of a cell can be precisely inherited by its offspring. The two functions of metabolism and replication are again well described by the metaphor of hardware and software. Metabolism is the business of protein, replication is the business of nucleic acid. Metabolism is a hardware function because it requires constant activity. Replication is a software function because it requires stability and legibility. Nucleic acids, like floppy disks, are easily read and copied. Proteins, like computers, are made by following instructions and not by copying.

The metaphor of the computer will serve as a framework within which to examine the problem of origins. I approach the discussion of the origin of life by introducing a selection of people who have contributed to our thinking. I have six characters to introduce. I might have titled this chapter as Luigi Pirandello titled his play, *Six Characters in Search of an Author*. The first two characters are the physicist Erwin Schrödinger and the mathematician John von Neumann.

In 1943, the year before Avery discovered the transformation of cells by nucleic acids, Erwin Schrödinger gave a course of lectures on biology to a mixed audience at Trinity College, Dublin. Schrödinger did not then know about Avery's experiment. Schrödinger did not once mention nucleic acids in his lectures. This makes it all the more remarkable that he was able to guess correctly the general nature and properties of the genetic apparatus in which nucleic acids play the dominant role.

The year 1943 was a bleak moment in the history of mankind. Ireland was then, as it had been in the days of St. Columba 1400 years earlier, a refuge for scholars and a nucleus of civilization beyond the reach of invading barbarians. It was one of the few places in Europe where peaceful scientific meditation was still possible. Schrödinger proudly remarks in the published version of his lectures that they were given "to an audience of about four hundred which did not substantially dwindle." The lectures were published in 1944, in a little book with the title *What Is Life?* Schrödinger's book is less than a hundred pages long. It was widely read, and was influential in guiding the thoughts of the young people who created the new science of molecular biology in the following decade. It is clearly and simply written, with only five references to the technical literature and less than ten equations from beginning to end.

Schrödinger's book was seminal because he knew how to ask the right questions. The basic questions which he asked were the following: What is the physical structure of the molecules which are duplicated when cells divide? How is the process of duplication to be understood? How do these molecules retain their individuality from generation to generation? How do they succeed in controlling the metabolism of cells? And how do they create the organization that is visible in the structure and function of higher organisms? He did not answer these questions. But by asking them he set biology moving along the path which led to the epochmaking discoveries of the subsequent forty years, to the double helix, to the genetic code, to the precise analysis and wholesale synthesis of genes, to the quantitative measurement of evolutionary divergence of species.

Schrödinger showed wisdom not only in the questions which he asked but also in the questions he did not ask. He did not ask any questions about the origin of life. He understood that the time was ripe in 1943 for a fundamental understanding of the physical basis of life. He also understood that the time was not then ripe for any fundamental understanding of life's origin. Until the basic chemistry of living processes was clarified, one could not ask meaningful questions about the possibility of spontaneous generation of these processes in a pre-biotic environment. He wisely left the question of origins to a later generation.

Now, forty years later, the time is ripe to ask the questions which Schrödinger avoided. The questions of origin are now becoming experimentally accessible, just as the questions of structure were becoming experimentally accessible in the 1940s. Schrödinger asked the right questions about structure because his thoughts were based on the experimental discoveries of his friend Max Delbrück. We can hope to ask the right questions about origins today because our thoughts are guided by the experimental discoveries of Manfred Eigen and Leslie Orgel. Delbrück became the chief explorer of the problems of structure in the 1930s and 1940s because he hit on the bacteriophage as the ideal experimental tool, a biological system stripped of inessential complications and reduced to an almost bare genetic apparatus. The phage is a virus specialized for living as a parasite inside a bacterium. The phage was for biology what the hydrogen atom was for physics. In a similar way, Eigen became the chief explorer of the problems of origin in the 1970s because he hit on RNA, one of the common varieties of nucleic acid, as the ideal experimental tool for studies of molecular evolution in the test tube. An RNA molecule is essentially a naked gene. Eigen's RNA experiments have carried Delbrück's phage experiments one step further, stripping the genetic apparatus completely naked and enabling us to study its replication unencumbered by the baggage of structural molecules which even so rudimentary a creature as a bacteriophage carries with it.

Before discussing the experiments of Eigen and Orgel in detail, I want to finish my argument with Schrödinger. I venture to say that Schrödinger in his discussion of the nature of life missed an essential point. In Schrödinger's book we find four chapters describing in lucid detail the phenomenon of biological replication, and a single chapter describing less lucidly the phenomenon of metabolism. He finds a conceptual basis in physics both for exact replication and for metabolism. Replication is explained by the quantum-mechanical stability of molecular structures, while metabolism is explained by the ability of a living cell to extract free energy from its surroundings in accordance with the laws of thermodynamics. Schrödinger was evidently more interested in replication than in metabolism. There are two obvious reasons for his bias. He was, after all, one of the inventors of quantum mechanics, and it was natural for him to be primarily concerned with the biological implications of his own brainchild. Secondly, his thinking was based on Delbrück's experiments, which were biased in the same direction. Delbrück's experimental system, the bacteriophage, is a purely parasitic creature in which the metabolic function has been lost and only the replicative function survives. It was indeed precisely this concentration of attention upon a degenerate and highly specialized form of life which enabled Delbrück to do experiments exploring the physical basis of biological replication. It was necessary to find a creature without metabolism in order to isolate experimentally the phenomena of replication. Delbrück penetrated more deeply than his contemporaries into the mechanics of replication because he was not distracted by the problems of metabolism. Schrödinger saw the world of biology through

Delbrück's eyes. It is not surprising that Schrödinger's view of a living organism resembles a bacteriophage more than it resembles a bacterium or a fruit-fly. His single chapter devoted to the metabolic aspect of life appears to be an afterthought, put in for the sake of completeness but not affecting the main line of his argument.

The main line of Schrödinger's argument, leading from the facts of biological replication to the quantum-mechanical structure of the gene, was brilliantly right and fruitful. It set the style for the subsequent development of molecular biology. Neither Schrödinger himself nor the biologists who followed his lead appear to have been disturbed by the logical gap between his main argument and his discussion of metabolism. Looking back on his 1943 lectures now with the benefit of forty years of hindsight, we may wonder why he did not ask some fundamental questions which the gap might have suggested to him. Is life one thing or two things? Is there a logical connection between metabolism and replication? Can we imagine metabolic life without replication, or replicative life without metabolism? These questions were not asked because Schrödinger and his successors took it for granted that the replicative aspect of life was primary, the metabolic aspect secondary. As their understanding of replication became more and more triumphantly complete, their lack of understanding of metabolism was pushed into the background. In popular accounts of molecular biology as it is now taught to schoolchildren, life and replication have become practically synonymous. In modern discussions of the origin of life it is often taken for granted that the origin of life was the same thing as the origin of replication. Manfred Eigen is an extreme example of this tendency. He chose as the working material for his experiments the substance RNA which is ideal for studying replication but incapable of metabolism. Eigen's theories of the origin of life are in fact theories of the origin of replication.

It is important here to make a sharp distinction between replication and reproduction. I shall be suggesting as a hypothesis that the earliest living creatures were able to reproduce but not to replicate. What does this mean? For a cell, to reproduce means simply to divide into two cells with the daughter cells inheriting approximately equal shares of the cellular constituents. For a molecule, to replicate means to construct a precise copy of itself by a chemically specific process. Cells can reproduce but only molecules can replicate. In modern times, reproduction of cells is always accompanied by replication of molecules, but this need not always have been so in the past. Conversely, even in modern cells, replication of molecules often occurs without reproduction of cells.

Only five years after Schrödinger gave his lectures in Dublin, the logical relations between replication and metabolism were clarified by the mathematician John von Neumann. In a lecture which he gave at Princeton in 1948, Von Neumann introduced the computer as a metaphor for a living cell. He described an analogy between the functioning of living organisms and the functioning of mechanical automata. His automata were an outgrowth of his thinking about computers. Out of his computers, Von Neumann built a grand conceptual scheme which he called "The General and Logical Theory of Automata." A Von Neumann automaton had two essential components which were later, when his ideas were taken over by the computer industry, given the names "hardware" and "software." Hardware processes information; software embodies information. In the old days, hardware was machinery and software was paper. Now hardware is silicon chips and software is floppy disks.

Von Neumann described precisely in abstract terms the logical connections between the components. For a complete self-reproducing automaton, both components are essential. But there is an important sense in which hardware comes logically prior to software. Pocket calculators existed before there were floppy disks. An automaton composed of hardware without software can exist and maintain its own metabolism. It can live independently for as long as it finds food to eat or numbers to crunch. An automaton composed of software without hardware must be an obligatory parasite. It can function only in a world already containing other automata whose hardware it can borrow. It can replicate itself only if it succeeds in finding a cooperative host automaton, just as a bacteriophage can replicate only if it succeeds in finding a cooperative bacterium.

Let me summarize the story up to this point. Our illustrious predecessor Erwin Schrödinger gave his book the title *What Is Life?* but neglected to ask whether the two basic functions of life, metabolism and replication, are separable or inseparable. Our illustrious predecessor John von Neumann raised the question which Schrödinger had missed and gave it a provisional answer. Von Neumann observed that metabolism and replication, however intricately they may be linked in the biological world as it now exists, are logically separable. It is logically possible to postulate organisms composed of pure hardware, capable of metabolism but incapable of replication. It is possible to postulate organisms composed of pure software, capable of replication but incapable of metabolism. And if the functions of life are separated in this fashion, it is to be expected that the latter type of organism will become an obligatory parasite upon the former. This logical analysis of the functions of life helps to explain and to correct the bias toward replication which is evident in Schrödinger's thinking and in the whole history of molecular biology. Organisms specializing in replication tend to be parasites, and molecular biologists prefer parasites for experimental study because parasites are structurally simpler than their hosts and better suited to quantitative manipulation. In the balance of nature there must be an opposite bias. Hosts must exist before there can be parasites. The survival of hosts is a precondition for the survival of parasites. Somebody must eat and grow to provide a home for those who only replicate. In the living world, as in the world of human society and economics, we cannot all be parasites.

When we begin to think about life's origins we meet again the question which Schrödinger did not ask, Is life one thing or two? And we meet again Von Neumann's answer, that life is two things, metabolism and replication, and that the two things are logically separable. There are accordingly two logical possibilities for life's origins. Either life began only once, with the functions of replication and metabolism already present in rudimentary form and linked together from the beginning. Or life began twice, with two separate kinds of creatures, one kind capable of metabolism without exact replication, the other kind capable of replication without metabolism. If life began twice, the first beginning must have been with proteins, the second beginning with nucleic acids. The first protein creatures might have existed independently for a long time, eating and growing and gradually evolving a more and more efficient metabolic apparatus. The nucleic acid creatures must have been obligatory parasites from the start, preying upon the protein creatures and using the products of protein metabolism to achieve their own replication.

The main theme of this and the next chapter will be a critical examination of the second possibility, the possibility that life began twice. I call this possibility the double-origin hypothesis. It is a hypothesis, not a theory. A theory of the origin of life should describe in some detail a postulated sequence of events. The hypothesis of dual origin is compatible with many theories. It may be useful to examine the consequences of the hypothesis without committing ourselves to any particular theory.

I do not claim that the double-origin hypothesis is true, or that it is supported by any experimental evidence. Indeed, my purpose is just the opposite. I would like to stimulate experimental chemists and biologists and paleontologists to find the evidence by which the hypothesis might be tested. If it can be tested and proved wrong, it will have served its purpose. We will then have a firmer foundation of fact on which to build theories of single origin. If the double-origin hypothesis can be tested and not proved wrong, we can proceed with greater confidence to build theories of double origin. The hypothesis is useful only insofar as it may suggest new experiments.

Lacking new experiments, we have no justification for believing strongly in either the single-origin or the double-origin hypothesis. I have to confess my own bias in favor of double origin. But my bias is based only on general philosophical preconceptions, and I am well aware that the history of science is strewn with the corpses of dead theories which were in their time supported by the prevailing philosophical viewpoints. For what it is worth, I may state my philosophical bias as follows. The most striking fact which we have learned about life as it now exists is the ubiquity of dual structure, the division of every organism into hardware and software components, into protein and nucleic acid. I consider dual structure to be prima facie evidence of dual origin. If we admit that the spontaneous emergence of protein structure and of nucleic acid structure out of molecular chaos are both unlikely, it is easier to imagine two unlikely events occurring separately over a long period of time than to imagine two unlikely events occurring simultaneously. Needless to say, vague arguments of this sort, invoking probabilities which we are unable to calculate quantitatively, cannot be conclusive.

The third and fourth on my list of illustrious predecessors are Manfred Eigen and Leslie Orgel. Unlike Schrödinger and Von Neumann, they are experimenters. They are the chief explorers of the experimental approaches to the problem of the origin of life. They are, after all, chemists, and this is a job for chemists. Eigen and his colleagues in Germany have done experiments which show us biological organization originating spontaneously and evolving in a test tube. Eigen won his Nobel Prize for work in the more conventional areas of physical chemistry. After winning the prize, he switched to the less conventional area of molecular evolution. As a result, he has now become the leading guru in the international fraternity of people who think seriously about the origin of life.

The Eigen experiments begin with a supply of the building blocks which make up RNA when they are joined together by the right chemical bonds. The building blocks are called nucleotides. They come in four kinds. Eigen dissolved these four kinds of molecule in water and mixed them together with a protein which catalyzes the replication of RNA. The experiment was done first with a small quantity of RNA added to the test tube. Then the protein replicates the RNA by collecting nucleotides from the solution and linking them together. The amount of RNA in the solution can be measured as a function of time and is observed to double every two or three minutes. The process of replication is very quick and efficient, once there is some RNA in the tube to be replicated. The experiment was done using smaller and smaller quantities of RNA as the initial seed. It turned out that it works perfectly well with only a single molecule of RNA. A single RNA molecule will be replicated enough times in an hour or two to make 10^{14} copies, at which point the supply of nucleotides begins to be exhausted. The replication of the RNA from one molecule to 10^{14} is reproducible and predictable.

But then Eigen tried something new. He did the experiment with zero molecules of RNA in the tube. He gave the nucleotides nothing to copy. You would expect then that nothing would happen. But in fact a lot of interesting things happened. The main thing that happened was that RNA appeared in the tube. But it did not appear reproducibly. It took typically two or three hours for an RNA molecule to appear, but the time varied from experiment to experiment. Once the first RNA molecule was produced by some statistically random process, it would then be replicated as before. So the experiment beginning with zero RNA ended as before with 10^{14} molecules of RNA, only the time taken to reach 10^{14} molecules was no longer predictable. Also, the sequence of nucleotides in the spontaneously generated RNA was not the same from run to run. In some cases, Eigen observed mutations in his RNA. The first RNA to appear would be replaced later by a mutated RNA which replicated itself more efficiently.

Eigen's experiments show that a solution of nucleotides will under suitable conditions give rise spontaneously to a nucleic acid molecule which replicates and mutates and competes with its progeny for survival. From a certain point of view, one might claim that these experiments already achieved the spontaneous generation of life from non-life. They bring us at least to the point where we can ask and answer questions about the ability of nucleic acids to synthesize and organize themselves. Unfortunately, the conditions in Eigen's test tubes are not really pre-biotic. To make his experiments work, Eigen put into the test tubes a protein catalyst extracted from a living bacteriophage. The synthesis and replication of the nucleic acid is dependent on the structural guidance provided by the protein. We are still far from an experimental demonstration of the appearance of biological order without the help of a biologically derived precursor. Nevertheless, Eigen has provided the tools with which we may begin to attack the problem of origins. He has brought the origin of life out of the domain of idle speculation and into the domain of experiment.

Leslie Orgel is, like Manfred Eigen, an experimental chemist. He taught me most of what I know about the chemical antecedents of life. He has done experiments complementary to those of Eigen. Eigen was able to make RNA grow out of nucleotides without providing any RNA for the nucleotides to copy, but with a protein catalyst to tell the nucleotides what to do. Orgel has done equally important experiments in the opposite direction, demonstrating that nucleotides will under certain conditions make RNA if they are given RNA molecules to copy, without any protein to guide them. Orgel found that zinc ions in the solution are a good catalyst for the RNA synthesis. It may not be entirely coincidental that many modern biological catalysts have zinc ions in their active sites. To summarize, Eigen made RNA using a protein but no RNA, and Orgel made RNA using a RNA but no protein. In living cells we make RNA using both RNA and protein. If we suppose that RNA was the original living molecule, then to understand the origin of life we have to make RNA using neither RNA nor protein. Neither Eigen nor Orgel has come close to achieving this goal.

The experiments of Eigen and Orgel fit more naturally into the framework of a double-origin hypothesis. According to the double-origin hypothesis, RNA was not the original living molecule. The original living molecules in this hypothesis were proteins, and life of a sort was already established before RNA came into the picture. The Eigen and Orgel experiments are exploring the evolution of RNA under conditions appropriate to the second origin of life. They come close to describing a parasitic development of RNA life within an environment created by a pre-existing protein life. Concerning the first origin of life, the origin of protein life and of protein metabolism, they say nothing. The origin of metabolism is the next great virgin territory which is waiting for the experimental chemists to explore.

The fifth on my list of illustrious predecessors is Lynn Margulis. She is an ecologist, a first-rate scientist and writer. But she also performed another service to the popular understanding of science. She was Carl Sagan's first wife and taught him most of what he knows about biology. Although she is still very much alive and considerably younger than I am, she set the style in which I came to think about early evolution.

Lynn Margulis is one of the chief bridge builders in modern biology. She built a bridge between the facts of cellular anatomy and the facts of molecular genetics. Her bridge was the idea that parasitism and symbiosis were the driving forces in the evolution of cellular complexity. She did not invent this idea, but she was its most active promoter and systematizer. She collected the evidence to support her view that most of the internal structures of cells did not originate within the cells but are descended from independent living creatures which invaded the cells from outside like carriers of an infectious disease. Her book *Symbiosis in Cell Evolution* summarizes the evidence up to 1981. According to Margulis, the invading creatures and their hosts gradually evolved into a relationship of mutual dependence, so that the erstwhile disease organism became by degrees a chronic parasite, a symbiotic partner, and finally an indispensable part of the substance of the host.

This Margulis picture of early cellular evolution now has incontrovertible experimental support. Some of the molecular structures within plant and animal cells are found to be related more closely to alien bacteria than to the cells in which they have been incorporated for 1 or 2 billion years. There are also reasons for believing that the Margulis picture will be valid even in cases where it cannot be experimentally demonstrated. A living cell, in order to survive, must be intensely conservative. It must have a finely tuned molecular organization and it must have efficient mechanisms for destroying promptly any molecules which depart from the overall plan. Any new structure arising within this environment must be an insult to the integrity of the cell. Almost by definition, a new structure will be a disease which the cell will do its best to resist. It is possible to imagine new structures arising internally within the cell and escaping its control, like a cancer growing in a higher organism. But it is easier to imagine new structures coming in from the outside like infectious bacteria, already prepared by the rigors of independent living to defend themselves against the cell's efforts to destroy them.

The main reason why I find the double-origin hypothesis congenial is that it fits well into the general picture of evolution portrayed by Margulis. According to Margulis, most of the big steps in cellular evolution were caused by parasites. The double-origin hypothesis implies that nucleic acids were the oldest and most successful cellular parasites. It extends the scope of the Margulis picture of evolution even further back into the past. It proposes that the original living creatures were cells with a metabolic apparatus but with no genetic apparatus. Such cells would lack the capacity for exact replication but could grow and divide and reproduce themselves in an approximate statistical fashion. They might have continued to exist for millions of years, gradually diversifying and refining their metabolic pathways. Amongst other things, they discovered how to synthesize ATP, adenosine triphosphate, the magic molecule which serves as the principal energy-carrying intermediate in all modern cells. Cells carrying ATP were able to function more efficiently and prevailed in the Darwinian struggle for existence. In time it happened that cells were full of ATP and other related molecules such as AMP, adenosine monophosphate, and so on.

Now we observe the strange fact that the two molecules ATP and AMP, having almost identical chemical structures, have totally different but equally essential functions in modern cells. ATP is hardware. AMP is software. ATP is the universal energy-carrier. AMP is one of the nucleotides which make up RNA and function as bits of information in the genetic apparatus. ATP is a compound composed of an adenine base, a sugar and three phosphate ions, joined together in a particular geometrical arrangement. AMP has the same pieces in the same arrangement except that two of the phosphate ions are missing. To get from ATP to AMP, all you have to do is replace a triple phosphate by a single phosphate.

I am proposing that the primitive cells had no genetic apparatus but were saturated with molecules like AMP as a result of the energy-carrying function of ATP. This was a dangerously explosive situation. In one cell which happened to be carrying an unusually rich supply of nucleotides, an accident occurred. The nucleotides began doing the Eigen experiment on RNA synthesis 3 billion years before it was done by Eigen. Within the cell, with some help from pre-existing proteins, the nucleotides produced an RNA molecule which then continued to replicate itself. In this way RNA first appeared as a parasitic disease within the cell. The first cells in which the RNA disease occurred probably became sick and died. But then, according to the Margulis scheme, some of the infected cells learned how to survive the infection. The protein-based life learned to tolerate the RNA-based life. The parasite became a symbiont. And then, very slowly over millions of years, the protein-based life learned to make use of the capacity for exact replication which the chemical structure of RNA provided. The primal symbiosis of protein-based life and parasitic RNA grew gradually into a harmonious unity, the modern genetic apparatus.

This view of RNA as the oldest and most incurable of our parasitic diseases is only a poetic fancy, not yet a serious scientific theory. Still, it is attractive to me for several reasons. First, it is in accordance with our human experience that hardware should come before software. The modern cell is like a computer-controlled chemical factory in which the proteins are the hardware and the nucleic acids are the software. In the evolution of machines and computers, we always developed the hardware first before we began to think about software. I find it reasonable that natural evolution should have followed the same pattern. A second argument in favor of the parasite theory of RNA comes from chemistry. Because of the details of the chemistry, it is easier to imagine a pond on the pre-biotic Earth becoming a rich soup of amino acids than to imagine a pond becoming a rich soup of nucleotides. Nucleotides would have had a better chance to survive if they originated in biological processes inside already existing cells. My third reason for liking the parasite theory of RNA is that it may be experimentally testable. If the theory is true, living cells may have existed for a very long time before becoming infected with nucleic acids. There exist microfossils, traces of primitive cells, in rocks which are more than 3 billion years old. It is possible that some of these microfossils might come from cells older than the origin of RNA. It is possible that the microfossils may still carry evidence of the chemical nature of the ancient cells. For example, if the microfossils were found to preserve in their mineral constituents significant quantities of phosphorus, this would be strong evidence that the ancient cells already possessed something resembling a modern genetic apparatus. So far as I know, no such evidence has been found. I do not know whether the processes of fossilization would be likely to leave chemical traces of nucleic acids intact. So long as this possibility exists, we have the opportunity to test the hypothesis of a late origin of RNA by direct observation.

The last of the illustrious predecessors on my list is the geneticist Motoo Kimura. Kimura lives in Japan and has a biological institute of his own out in the country between

Tokyo and Kyoto, with a magnificent view of Mount Fujiyama. Kimura developed the mathematical basis for a statistical treatment of molecular evolution, and he has been the chief advocate of the neutral theory of evolution. The neutral theory says that, through the history of life from beginning to end, random statistical fluctuations have been more important than Darwinian selection in causing species to evolve. Evolution by random statistical fluctuation is called genetic drift. Kimura says that genetic drift drives evolution more powerfully than natural selection.

I am indebted to Kimura in two separate ways. First, I use Kimura's mathematics as a tool for calculating the behavior of molecular populations. The mathematics is correct and useful, whether you believe in the neutral theory of evolution or not. Second, I find the neutral theory helpful even though I do not accept it as dogma. In my opinion, Kimura has overstated his case, but still his picture of evolution may sometimes be right. Genetic drift and natural selection are both important, and there are times and places where one or the other may be dominant. In particular, I find it reasonable to suppose that genetic drift was dominant in the very earliest phase of biological evolution, before the mechanisms of heredity had become exact. Even if the neutral theory is not true in general, it may be a useful approximation to make in building models of pre-biotic evolution.

We know almost nothing about the origin of life. We do not even know whether the origin was gradual or sudden. It might have been a process of slow growth stretched out over millions of years, or it might have been a single molecular event that happened in a fraction of a second. As a rule, natural selection is more important over long periods of time and genetic drift is more important over short periods. If you think of the origin of life as slow, you must think of it as a Darwinian process driven by natural selection. If you think of it as quick, the Kimura picture of evolution by statistical fluctuation without selection is appropriate. In reality the origin of life must have been a complicated process, with incidents of rapid change separated by long periods of slow adaptation. A complete description needs to take into account both drift and selection. If one wishes to examine seriously the double-origin hypothesis, the hypothesis that life began and flourished without the benefit of exact replication, then it is natural to imagine that genetic drift remained strong and natural selection remained relatively weak during the early exploratory phases of evolution.

There are many other illustrious predecessors besides those whom I have mentioned. I chose to speak of these six—Schrödinger, Von Neumann, Eigen, Orgel, Margulis, and Kimura—because each of them is in some sense a philosopher as well as a scientist. Each of them brought to biology not just technical skills and knowledge but a personal philosophical viewpoint extending beyond biology over the whole of science. From all of them I have borrowed the ideas which fitted together to form my own viewpoint. The origin of life is one of the few scientific problems which are broad enough to make use of ideas from almost all scientific disciplines. Schrödinger brought to it ideas from physics, Von Neumann from mathematical logic, Eigen and Orgel from chemistry, Margulis from ecology, and Kimura from population biology. I will try in the next chapter to explore the connections, to see whether mathematical logic and population biology may have raised new questions which chemistry may be able to answer.

5

WHY IS LIFE SO COMPLICATED?

The origin of life, however it happened, must have been the culminating event at the end of a long and complicated process of preparation. The process by which the planet Earth prepared itself for life is called pre-biotic evolution. The study of pre-biotic evolution divides itself into three main stages which one may label with the names "geophysical," "chemical" and "biological." The geophysical stage concerns itself with the early history of the earth and especially with the nature of the Earth's primitive atmosphere. The chemical stage concerns itself with the synthesis, by natural processes operating within plausible models of the primitive atmosphere and ocean, of the chemical building blocks of life. By building blocks we mean principally the amino acids and nucleotides out of which proteins and nucleic acids are built. The biological stage concerns itself with the appearance of biological organization, with the building of a coordinated population of proteins and nucleic acids and other large molecules out of a random assortment of building blocks.

The geophysical and chemical stages of pre-biotic evolution are reasonably well understood. These two stages are in the hands of competent experts, and I have nothing significant to add to what the experts have reported. Theories of the geophysical stage can be checked by abundant observations in the fields of geochemistry, meteoritic chemistry and radio astronomy. Theories of the chemical stage can be checked by experiments done by chemists in the laboratory. Many details remain to be elucidated, but the geophysical and chemical stages are no longer shrouded in mystery. I have therefore concentrated my attention on the biological stage. The problem of the origin of life is for me the biological stage, the problem of the appearance of biological organization out of molecular chaos. It is in this biological stage that mysteries still remain. The purpose of my own work has been to try to describe mathematically what we mean by the appearance of biological organization, and thereby to make the biological stage accessible to experimental study.

It is likely that there existed on the primitive Earth substantial quantities of simple organic molecules which would have either dissolved in the primitive oceans or formed an oily scum on their surface. Molecules of this sort are seen by radio astronomers in the sky and by geochemists in carbonaceous meteorites. The geophysical stage of pre-biotic evolution thus provides us with a wide choice of hydrogen-rich gases and liquids to serve as the input for the chemical stage.

The chemical stage of pre-biotic evolution was explored in the classic experiments of Stanley Miller in 1953, and in many later experiments. Miller took a plausible reducing atmosphere composed of methane, ammonia, molecular hydrogen and water, passed electric sparks through it, and collected the reaction products. He found a mixture of organic compounds containing a remarkably high fraction of amino acids. Other people have repeated the Miller experiments with many variations, using ultraviolet light or ionizing radiation as the energy source instead of electric sparks. The results are consistent. The upshot of the Miller experiments is that we can rely on nature to provide an ample supply of amino acids on the primitive Earth. So long as there was a reducing atmosphere with plenty of ultraviolet light pouring in from the Sun, amino acids must have been made in large quantities. They would have descended from the atmosphere with rainfall and would have accumulated in land-locked lakes and seas where rainwater evaporates and leaves its dissolved chemicals behind. This is a familiar story, ending with a picture of a pond full of rich warm soup on the lifeless Earth. The pond contains a concentrated solution of amino acids and other organic molecules, waiting only for the breath of life to stir them into organized activity. So far as the geophysical and chemical stages are concerned, this story is probably a good approximation to the truth.

The pre-biotic synthesis of nucleotides, the pieces out of which nucleic acids are made, is a more difficult problem. Efforts to synthesize nucleotides directly from their elementary components in a Miller-style experiment have not been successful. A nucleotide is a wobblier and more delicate structure than an amino acid. The nucleotides on the primitive Earth would have been rare birds, difficult to synthesize and easy to dissociate. Nobody has yet discovered a way to make them out of their components rapidly enough so that they would have a reasonable chance of finding each other and combining into nucleic acids before they fall apart.

The results of thirty years of intensive chemical experimentation have shown that the pre-biotic synthesis of amino acids is easy to simulate but the pre-biotic synthesis of nucleotides is not. We cannot say that the pre-biotic synthesis of nucleotides is impossible. We know only that if it happened, it happened by some process which none of our chemists has been clever enough to reproduce. This conclusion may be considered to favor the double-origin hypothesis and to argue against a single-origin hypothesis for the origin of life. A single-origin hypothesis requires both amino acids and nucleotides to be synthesized by natural processes before life began. The double-origin hypothesis requires only amino acids to be synthesized pre-biotically, the nucleotides being formed later as a by-product of protein metabolism. The evidence from chemical simulations does not disprove the single-origin hypothesis, but makes at least a strong presumptive case against it.

We have experimental and observational evidence concerning things which happened before and after the origin of life. Before the origin of life, we had geophysical and chemical processes which left behind traces that can be observed in the Earth's rocks and in the sky. After the origin of life, we had evolutionary processes which can be observed in fossils. Concerning the origin of life itself, the watershed between chemistry and biology, the transition between lifeless organic soup and organized biological metabolism, we have no direct evidence at all. The crucial transition from disorder to order left behind no observable traces. When we try to understand the nature of this transition, we are forced to go beyond experimental evidence and take refuge in theory.

There are three main groups of theories about the origin of life. I call them after the names of their most famous advocates, Oparin, Eigen, and Cairns-Smith. I have not done the historical research that would be needed to find out who thought of them first. The Oparin theory was described in Alexander Oparin's book *The Origin of Life* in 1924, long before anything was known about the structure and chemical nature of genes. Oparin supposed that the order of events in the origin of life was: cells first, proteins second, genes third. He observed that when a suitably oily liquid is mixed with water, it sometimes happens that the two liquids form a stable mixture called a coacervate, with the oily liquid dispersed into small droplets which remain suspended in the water. Coacervate droplets are easily formed by non-biological processes, and they have a certain superficial resemblance to living cells. Oparin proposed that life began by the successive accumulation of more and more complicated molecular populations within the droplets of a coacervate. The physical framework of the cell came first, provided by the naturally occurring droplet. The proteins came second, organizing the random population of molecules within the droplet into self-sustaining metabolic cycles. The genes came third, since Oparin had only a vague idea of their function and they appeared to him to belong to a higher level of biological organization than proteins.

The Oparin picture was generally accepted by biologists for half a century. It was popular, not because there was any evidence to support it, but rather because it seemed to be the only alternative to biblical creationism. Then, during the last twenty years, Manfred Eigen provided another alternative by turning the Oparin theory upside down. The Eigen theory reverses the order of events, placing genes first, proteins second, and cells third. It is now the most fashionable and generally accepted theory. It has become popular for two reasons. First, Avery's discovery that genes are nucleic acids made genes easier to study than proteins. Second, the experiments of Eigen and of Orgel use RNA as working material and make it plausible that the replication of RNA was the fundamental process around which the rest of biology developed. Once the mystery of the genetic code was understood, it became natural to think of the nucleic acids as primary and of the proteins as secondary structures. Eigen's theory has self-replicating RNA at the beginning, proteins appearing soon afterwards to build with the RNA a primitive form of the modern genetic apparatus, and cells appearing later to give the apparatus physical cohesion.

The third theory of the origin of life, the theory of Graham Cairns-Smith, is based upon the idea that naturally occurring microscopic crystals of the minerals contained in common clay might have served as the original genetic material before nucleic acids were invented. The microcrystals of clay consist of a regular lattice with a regular array of atomic sites, but with an irregular distribution of metals such as magnesium and aluminum occupying the sites. The metal atoms can be considered as carriers of information like the nucleotide bases in a molecule of RNA. A microcrystal of clay is usually a flat plate with two plane surfaces exposed to the surrounding medium. Suppose that a microcrystal is contained in a droplet of water with a variety of organic molecules dissolved in the water. The metal atoms embedded in the plane surfaces form irregular patterns which can adsorb particular molecules to the surfaces and catalyze chemical reactions on the surfaces in ways dependent on the precise arrangement of the atoms.

In this fashion the information contained in the pattern of atoms might be transferred to chemical species dissolved in the water. The crystal might perform the same function as RNA in guiding the metabolism of amino acids and proteins. Moreover, it is conceivable that the clay microcrystal can also replicate the information contained in its atoms. When the crystal grows by accreting dissolved atoms from the surrounding water, the newly accreted layer will tend to carry the same pattern of metals as the layer below it. If the crystal is later cut along the plane separating the old from the new material, we will have a new exposed surface replicating the original pattern. The clay crystal is thus capable in principle of performing both of the essential functions of a genetic material. It can replicate the information which it carries, and it can transfer the information to other molecules. It can do these things in principle. That is to say, it can do them with some undetermined efficiency which may be very low. There is no experimental evidence to support the statement that clay can act either as a catalyst or as a replicator with enough specificity to serve as a basis for life. Cairns-Smith asserts that the chemical specificity of clay is adequate for these purposes. The experiments to prove him right or wrong have not been done.

The Cairns-Smith theory of the origin of life has clay first, proteins second, cells third, and genes fourth. The beginning of life was a natural clay crystal directing the synthesis of protein molecules adsorbed to its surface. Later, the clay and the proteins learned to make cell membranes and became encapsulated in cells. The cells contained clay crystals performing in a crude fashion the functions performed in a modern cell by nucleic acids. This primeval clay-based life may have existed and may have evolved for many millions of years. Then one day a cell made the discovery that RNA is a better genetic material than clay. As soon as RNA was invented, the cells using RNA had an enormous advantage in metabolic precision over the cells using clay. The clay-based life was eaten or squeezed out of existence and only the RNA-based life survived.

At the present time there is no compelling reason to accept or to reject any of the three theories. Any of them, or none of them, could turn out to be right. We do not yet know how to design experiments which might decide between them. I happen to prefer the Oparin theory, not because I think it is necessarily right but because it is unfashionable. In recent years the attention of the experts has been concentrated upon the Eigen theory, and the Oparin theory has been neglected. The Oparin theory deserves a more careful analysis in the light of modern knowledge. My own work on the origin of life grew out of an attempt to put the Oparin theory into a modern framework using the mathematical methods of Kimura.

I find it illuminating to look at these theories in the light of the question which I raised in Chapter 4, whether the origin of life was single or double, whether metabolism and replication originated together or separately. The Cairns-Smith theory is explicitly a double-origin theory. It has the first origin of life mainly concerned with the building of a protein metabolic apparatus, the clay particles adding to this apparatus a replicative element which may or may not be essential. The second origin of life, which Cairns-Smith calls "genetic take-over," is the replacement of the clay component by an efficient replicative apparatus made of nucleic acids. Cairns-Smith imagines the two origins of life to be separated by a long period of biochemical evolution, so that the nucleic acid invasion occurs in cells already highly organized with proteins and membranes. The Oparin theory and the Eigen theory were presented as single-origin theories. Each of them supposes the origin of life to have been a single process. Oparin places primary emphasis on metabolism and barely discusses replication. Eigen places primary emphasis on replication and imagines metabolism falling into place rapidly as soon as replication is established. I am suggesting that the Oparin and Eigen theories make more sense if they are put together and interpreted as the two halves of a double-origin theory. In this case, Oparin and Eigen may both be right. Oparin is describing the first origin of life and Eigen the second. With this interpretation, we combine the advantages of the two theories and eliminate their most serious weaknesses. Moreover, the combination of Oparin and Eigen into a double-origin theory is not very different from the theory of Cairns-Smith. Roughly speaking, Cairns-Smith equals Oparin plus Eigen plus a little bit of clay. All three theories may turn out to contain essential elements of the truth.

There is a serious difficulty with the Eigen theory. The central problem for any theory of the origin of replication is the fact that a replicative apparatus has to function almost perfectly if it is to function at all. If it does not function perfectly, it will give rise to errors in replicating itself, and the errors will accumulate from generation to generation. The accumulation of errors will result in a progressive deterioration of the system until it is totally disorganized. This deterioration of the replicative apparatus is called the "error catastrophe." The same difficulty arises also in the Cairns-Smith theory. Since Cairns-Smith is relying on clay crystals to transmit genetic information from generation to generation, the copying of the crystals must be extraordinarily exact if the information is to be preserved.

Both in the Eigen theory and in the Cairns-Smith theory, the only way in which we can avoid the error catastrophe is to assume that there is a strong Darwinian selection process acting on the replicating molecules. We must assume that the errors in each generation are weeded out by selection. In other words, the correct molecules of RNA or clay must be rewarded and the incorrect molecules penalized by different rates of survival. Even when a strong Darwinian selection of this sort is operating, the error catastrophe will still occur unless the error rate is kept low. Eigen has worked out a mathematical theory of the error catastrophe. The conclusions of the theory are roughly as follows. Suppose that a self-replicating system is composed of N independent units. For example, in the Eigen theory the system is a molecule of RNA composed of N nucleotides, while in the Cairns-Smith theory the system is a clay crystal with N metal atoms on its surface. Then the error catastrophe will occur unless the error rate in the copying process is of the order of $(1/N)$ or smaller.

This condition for avoiding the error catastrophe is very stringent. It is barely satisfied in modern higher organisms which have N of the order of 10^8 and error rates of the order of 10^{-8}. To achieve an error rate as low as 10^{-8}, the modern organisms have evolved an extremely elaborate system of double-checking and error-correcting within the replication system. Before this delicate apparatus existed, the error rates must have been much higher. Eigen's criterion thus imposes severe requirements on any theory of the origin of life which makes replication a central element of life from the beginning.

All the experiments which have been done with RNA replication under abiotic conditions give error rates of the order of 10^{-2} at best. If we try to satisfy Eigen's criterion without the help of pre-existing organisms, we are limited to a replication system which can describe itself with less than one hundred bits of information. One hundred bits of information is far too few to describe any interesting protein chemistry. This does not mean that Eigen's theory is untenable. It means that Eigen's theory requires an information-processing system which is at the same time extraordinarily simple and extraordinarily free from error. We do not know how to achieve such low error rates in the initial phases of life's evolution.

I chose to study the Oparin theory because it offers a possible way of escape from the error catastrophe. In the Oparin theory, the origin of life is separated from the origin of replication. The first living cells had no system of precise replication and could therefore tolerate high error rates. The main advantage of the Oparin theory is that it allows early evolution to proceed in spite of high error rates. It has the first living creatures consisting of populations of molecules with a loose organization and no genetic fine-tuning. There is a high tolerance for errors because the metabolism of the population depends only on the catalytic activity of a majority of the molecules. The system can still function with a substantial minority of ineffective or uncooperative molecules. There is no requirement for unanimity. Since the statistical fluctuations in the molecular populations will be large, there is a maximum opportunity for genetic drift to act as driving force of evolution.

During the last few years I have been amusing myself with a simple theoretical model of the origin of life. I am trying to place the Oparin theory within a framework of strict mathematics so that its consequences can be calculated. I reduce the theory to a toy model by assuming a simple and arbitrary rule for the probability of molecular interactions. The entire intricate web of biochemical processes is replaced in the model by a single formula. The habit of constructing toy models of this sort is one to which theoretical physicists easily become addicted. When the real world is recalcitrant, we build ourselves toy models in which the equations are simple enough for us to solve. Sometimes the behavior of the toy model provides illuminating insight into the behavior of the real world. More often, the toy model remains what its name implies, a plaything for mathematically inclined physicists. In the present case, the toy model may have some connection with reality or it may not. Whether or not its premises are reasonable, at least its conclusions are definite. Given the premises from which it starts, it behaves as one would wish a primeval molecular population to behave, jumping with calculable probability between two states which differ by the presence or absence of metabolic organization.

We have here a situation analogous to the distinction between life and death in biological systems. I call the ordered state of a population of molecules "alive," since it has most of the molecules working together in a collaborative fashion to maintain the catalytic cycles which keep them active. I call the disordered state "dead," since it has the molecules uncoordinated and mostly inactive. A population, either in the dead or in the alive state, will generally stay there for a long time, making only small random fluctuations around the stable equilibrium. However, the population of molecules in a cell is finite, and there is always the possibility of a large statistical fluctuation which takes the whole population together over the saddle-point from one stable equilibrium to the other. When a "live" cell makes the big statistical jump over the saddle-point to the lower state, we call the jump "death." When a "dead" cell makes the jump up over the saddle-point to the upper state, we call the jump "origin of life."

There are three main ingredients in the model. First, elementary equations of population biology applied to a population of molecules. Second, elementary chemical thermodynamics giving a formula for the rates of various chemical reactions. Third, a statistical treatment of the transition from chemical disorder to chemical order, similar to the treatment of order-disorder transitions in physics.

We have then a definite model to work with. It remains to calculate its consequences, and to examine whether it shows interesting behavior. "Interesting behavior" here means the occurrence with reasonable probability of a jump from the disordered to the ordered state. We find that interesting behavior occurs for populations of molecules containing a total of about 10,000 building blocks. A population of several thousand building blocks linked into a few hundred large molecules would give a sufficient variety of structures to allow interesting catalytic cycles to exist. A population of the order of 10,000 is large enough to display the chemical complexity characteristic of life, and still small enough to allow the statistical jump from disorder to order to occur with probabilities which are not impossibly small.

The basic reason for the success of the model is its ability to tolerate high error rates. It overcomes the error catastrophe by abandoning exact replication. It neither needs nor achieves precise control of its molecular structures. It is this lack of precision which allows a population of 10,000 building blocks to jump into an ordered state without invoking a miracle. In a model of the origin of life which assumes exact replication from the beginning, with a low tolerance of errors, a jump of a population of N building blocks from disorder to order will occur spontaneously only with N of the order of 100 or less. In contrast, a non-replicating model can make the transition to order with a population a hundred times larger. The error rate in the ordered state of our model is typically between 20 and 30 percent. An error rate of 25 percent means that three out of four of the building blocks in each molecule are correctly placed. A catalyst with five building blocks in its active site has one chance out of four of being completely functional. Such a level of performance is tolerable for a non-replicating system, but would be totally unacceptable in a replicating system. The ability to function with a 25 percent error rate is the decisive factor which makes the ordered state in our model statistically accessible, with populations large enough to be biologically interesting.

One striking feature of our model which is absent in modern organisms is the symmetry between life and death. In the model, the ordered state and the disordered state are mirror images of each other. The probability of a transition from disorder to order is exactly equal to the probability of a transition from order to disorder. Death and resurrection occur with equal frequency. The origin of life is as commonplace an event as death.

How did it happen that, as life evolved, death continued to be commonplace while resurrection became rare? What happened was that the catalytic processes in the cell became increasingly fine-tuned and increasingly intolerant of error. The error rate in the metabolic apparatus of a modern cell is about 10^{-4}. An environmental insult such as a dose of X-rays which increases the error rate to 10^{-3} will disrupt the fine-tuned apparatus and cause the cell to die. Death is easy and resurrection is difficult, because the lethal error rate has moved so close to the ordered state and so far from the disordered state. For life to originate spontaneously it was essential to have an ordered state with a high error rate, but when life was once established the whole course of evolution was toward more specialized structures with lower tolerance of errors.

The model has many other unrealistic features. It includes none of the subtleties of structure and design that are present even in the simplest forms of life today. Nevertheless, it may throw some light on one of the central questions of biology, Why is life so complicated? The essential characteristic of living cells is homeostasis, the ability to maintain a steady and more or less constant chemical balance in a changing environment. Homeostasis is the machinery of chemical controls and feedback cycles which make sure that each molecular species in a cell is produced in the right proportion, not too much and not too little. Without homeostasis, there can be no ordered metabolism and no quasi-stationary equilibrium deserving the name of life. The question, Why is life so complicated? means in this context, Given that a population of molecules is able to maintain itself in homeostatic equilibrium at a steady level of metabolism, how many different molecular species must the population contain?

The biological evidence gives a rather definite answer to the question how many kinds of molecule are needed to make a homeostatic system, at least so long as we are talking about homeostatic systems of the modern type. There is a large number of different varieties of bacteria, and most of them contain a few thousand molecular species. If modern cells require a few thousand types of molecule for stable homeostasis, what does this tell us about primitive cells? Strictly speaking, it tells us nothing. Without the modern apparatus of genes, the ancient mechanisms of homeostasis must have been very different. In my toy model of the Oparin theory, the arithmetic of the model implies that the population of a cell at the origin of life should have been about 10,000 building blocks combined into a few hundred molecular species. The question remains, whether ancient cells in the real world could have achieved homeostasis with a few hundred molecules instead of the thousands required by modern cells. This question must be answered before we can build credible theories of the origin of life. It can be answered only by experiment. Forty years ago, Erwin Schrödinger suggested to biologists that they should investigate experimentally the molecular structure of the gene. That suggestion turned out to be timely. I am now suggesting that biologists investigate experimentally the population structure of homeostatic systems of molecules. If I am lucky, this suggestion may also turn out to be timely.

I now return to the toy model and examine some other questions which it raises. The questions are not specific to this particular model. They will arise for any model of the origin of life in which we have molecular populations achieving metabolism and homeostasis before they achieve replication. I comment briefly on each question in turn. After another twenty years of progress in biological research we may perhaps know whether my tentative answers are correct.

1. Was the Central Dogma of molecular biology true from the beginning?

The Central Dogma says that genetic information is carried only by nucleic acids and not by proteins. The dogma is true for all contemporary organisms, with one or two possible exceptions. The model implies that the dogma was untrue for the earliest forms of life. According to the model, the first cells passed genetic information to their offspring in the form of proteins. There is no logical reason why a population of proteins mutually catalyzing each other's synthesis should not serve as a carrier of genetic information.

The question, how much genetic information can be carried by a population of molecules without exact replication, is intimately bound up with the question of the nature of homeostasis. Homeostasis is the preservation of the chemical architecture of a population in spite of variations in local conditions and in the numbers of molecules of various kinds. Genetic information is carried in the architecture and not in the individual components. But we do not know how to define architecture or how to quantify homeostasis. Lacking a deep understanding of homeostasis, we have no way to calculate how many items of genetic information the homeostatic machinery of a cell may be able to preserve.

It seems to be true, both in the world of cellular chemistry and in the world of ecology, that homeostatic mechanisms have a general tendency to become complicated rather than simple. Homeostasis seems to work better with an elaborate web of interlocking cycles than with a small number of cycles working separately. Why this is so we do not know. We are back again with the question, Why is life so complicated? But the prevalence of highly complex homeostatic systems, whether we understand the reasons for it or not, is a fact. This fact is evidence that large amounts of genetic information can in principle be expressed in the architecture of molecular populations without nucleic acid software and without apparatus for exact replication.

2. How late was the latest common ancestor of all living species?

The evidence of similarity in the chemical architecture of modern species proves that all existing cells have a common ancestor but does not provide an absolute date for the epoch of the latest common ancestor. Genetic evidence gives us good relative dating of the different branches of the evolutionary tree, but no absolute dating. For absolute dates we must turn to the evidence of fossils. The pioneer in the discovery of fossil evidence for the absolute dating of early life was Elso Barghoorn. The rock in which microfossils are best preserved is chert, the geologists' name for the fine-grained silica rock which ordinary people call flint. Chert is formed by the slow precipitation of dissolved silica from water, a process which puts minimal stress on any small creatures that may become embedded within it. The chert once formed is hard and chemically inert, so that fossils inside it are well protected. The microfossils which Barghoorn and others have collected are little black blobs in which internal structure is barely discernible. Not all of them are definitely known to be organic in origin. I myself cannot pretend to decide whether a microscopic blob is a fossil cell or an ordinary grain of dust. I accept the verdict of the experts who say that most of the blobs are in fact fossils.

The results of a great number of observations of microfossils can be briefly summarized as follows. I use the word "eon" to mean a billion years. In rocks which are reliably dated with an age of about three eons, mainly in South Africa, we find microfossils which resemble modern bacteria in shape and size. In rocks which are dated with an age of about two eons, mainly in Canada, we find fossils which resemble modern algae, including chains of cells and other multicellular structures. In rocks which are dated with an age of one eon, mainly in Australia, we find fossils which resemble modern cells with some traces of internal structure.

The geological dating of the various fossil groups is accurate and reliable. Unfortunately, we do not know with equal accuracy what it is that is being dated. We do not know how to identify the various fossils with particular branches of the evolutionary tree. Except for the general similarity of size and shape, there is no evidence that the cells of the three-eon age group were cousins of modern bacteria. There is no evidence that they possessed a modern genetic apparatus. There is no evidence of the presence of nucleic acids in any of the ancient microfossils. So far as the evidence goes, the cells of the three-eon age group may have been either cells of modern type with a complete genetic apparatus, or cells of a rudimentary kind lacking nucleic acids altogether, or anything in between. Only the cells of the one-eon group were definitely modern. So far as the evidence goes, the latest common ancestor of all living cells may have lived before the three-eon group of fossils, or between the three-eon group and the two-eon group, or even later than the two-eon group. The dating of the latest common ancestor requires a reliable linkage of the branch points of the evolutionary tree with the various groups of fossils. The most urgent problem for evolutionary geneticists is to establish the calibration of relative ages determined by genetic linkages in terms of absolute ages determined by geology. Until this problem is solved, neither the genetic evidence nor the fossil evidence will be sufficient to determine the date of our latest common ancestor.

It is possible that the evolution of the modern genetic apparatus took eons to complete. The ancient microfossils may date from a time before there were genes. The pace of evolution may have accelerated after the genetic code was established. It is possible that the latest common ancestor came late in the history of life, perhaps as late as halfway from the beginning.

3. *Does there exist a chemical realization of my model, for example a population of a few thousand amino acids forming an association of proteins which can catalyze each other's synthesis with 80 percent accuracy? Can such an association of molecules be confined in a droplet and supplied with energy and raw materials in such a way as to maintain itself in a stable homeostatic equilibrium?*

These are the crucial questions which only experiment can answer.

4. *What will happen to my little toy model when the problem of the origin of life is finally solved?*

This is the last question raised by the model and it is easily answered. The answer was given nearly two hundred years ago by my favorite poet, William Blake:

To be an Error and to be Cast out is a part of God's Design.

At the end of his book *What Is Life?*, Schrödinger put a four-page epilogue with the title "On Determinism and Free Will." He there describes his personal philosophy, his way of reconciling his objective understanding of the physical machinery of life with his subjective experience of free will. Schrödinger writes with a clarity and economy of language which have rarely been equaled. I will not try to compete with him in summing up in four pages the fruits of a lifetime of philosophical reflection. Instead, I will end my discussion of the origin of life with some remarks about its wider implications, not for personal philosophy but for other areas of science. I use the word "science" here in a broad sense, including social as well as natural sciences. The sciences which I have particularly in mind are ecology, economics, and cultural history. In all these areas we are confronting the same question which is at the root of the problem of understanding the origin of life: Why is life so complicated? It may be that each of these areas has something to learn from the others.

The concept of homeostasis can be transferred without difficulty from molecular to ecological, economic and cultural contexts. In each area we have the unexplained fact that complicated homeostatic mechanisms are more prevalent, and seem to be more effective, than simple ones. This is most spectacularly true in the domain of ecology, where a typical stable community, for example, a few acres of woodland or a few square feet of grassland, comprises thousands of diverse species with highly specialized and interdependent functions. But a similar phenomenon is visible in economic life and in cultural evolution. The open-market economy and the culturally open society, notwithstanding all their failures and deficiencies, seem to possess a robustness which centrally planned economies and culturally closed societies lack. The homeostasis provided by unified five-year economic plans and by unified political control of culture does not lead to a greater stability of economies and cultures. On the contrary, the simple homeostatic mechanisms of central control have generally proved more brittle and less able to cope with historical shocks than the complex homeostatic mechanisms of the open market and the uncensored press.

But this chapter is not intended to be a political pep talk in defense of free enterprise. My purpose in mentioning the analogies between cellular and social homeostasis was not to draw a political moral from biology but rather to draw a biological moral from ecology and social history. Fortunately,

I can claim the highest scientific authority for drawing the moral in this direction. It is well known to historians of science that Charles Darwin was strongly influenced in his working out of the theory of evolution by his readings of the political economists from Adam Smith to Malthus and McCullough. Darwin himself said of his theory: "This is the doctrine of Malthus applied to the whole animal and vegetable kingdom." What I am proposing is to apply in the same spirit the doctrines of modern ecology to the molecular processes within a primitive cell.

In our present state of ignorance, we have a choice between two contrasting images to represent our view of the possible structure of a creature newly emerged at the first threshold of life. One image is the replicator model of Eigen, a molecular structure tightly linked and centrally controlled, replicating itself with considerable precision, achieving homeostasis by strict adherence to a rigid pattern. The other image is the "tangled bank" of Darwin, an image which Darwin put at the end of his *Origin of Species* to make vivid his answer to the question, What is Life?, an image of grasses and flowers and bees and butterflies growing in tangled profusion without any discernible pattern, achieving homeostasis by means of a web of interdependences too complicated for us to unravel.

The tangled bank is the image which I have in mind when I try to imagine what a primeval cell would look like. I imagine a collection of molecular species, tangled and interlocking like the plants and insects in Darwin's microcosm. This was the image which led me to think of error tolerance as the primary requirement for a model of a molecular population taking its first faltering steps toward life. Error tolerance is the hallmark of natural ecological communities, of free market economies and of open societies. I believe it must have been a primary quality of life from the very beginning. But replication and error tolerance are naturally antagonistic principles. That is why I like to exclude replication from the beginnings of life, to imagine the first cells as error-tolerant tangles of non-replicating molecules, and to introduce replication as an alien parasitic intrusion at a later stage. Only after the alien intruder has been tamed, the reconciliation between replication and error tolerance is achieved in a higher synthesis, through the evolution of the genetic code and the modern genetic apparatus.

The modern synthesis reconciles replication with error tolerance by establishing the division of labor between hardware and software, between the genetic apparatus and the gene. In the modern cell, the hardware of the genetic apparatus is rigidly controlled and error-intolerant. The hardware must be error-intolerant in order to maintain the accuracy of replication. But the error tolerance which I like to believe inherent in life from its earliest beginnings has not been lost. The burden of error tolerance has merely been transferred to the software. In the modern cell, with the infrastructure of hardware firmly in place and subject to a strict regime of quality control, the software is free to wander, to make mistakes and occasionally to be creative. The transfer of architectural design from hardware to software allowed the molecular architects to work with a freedom and creativity which their ancestors before the transfer could never have approached.

The analogies between the genetic evolution of biological species and the cultural evolution of human societies have been brilliantly explored by Richard Dawkins in his book *The Selfish Gene*. The book is mainly concerned with biological evolution; the cultural analogies are only pursued in the last chapter. Dawkins's main theme is the tyranny which the rigid demands of the replication apparatus have imposed upon all biological species throughout evolutionary history. Every species is the prisoner of its genes and is compelled to develop and to behave in such a way as to maximize their chances of survival. Only the genes are free to experiment with new patterns of behavior. Individual organisms must do what their genes dictate. This tyranny of the genes has lasted for 3 billion years and has only been precariously overthrown in the last hundred thousand years by a single species, Homo sapiens.

We have overthrown the tyranny by inventing symbolic language and culture. Our behavior patterns are now to a great extent culturally rather than genetically determined. We can choose to keep a defective gene in circulation because our culture tells us not to let hemophiliac children die. We have stolen back from our genes the freedom to make choices and to make mistakes.

In his last chapter Dawkins describes a new tyrant which has arisen within human culture to take the place of the old. The new tyrant is the "meme," the cultural analogue of the gene. A meme is a behavioral pattern which replicates itself by cultural transfer from individual to individual instead of by biological inheritance. Examples of memes are religious beliefs, linguistic idioms, fashions in art and science, in food and in clothes. Almost all the phenomena of evolutionary genetics and speciation have their analogues in cultural history, with the meme taking over the functions of the gene. The meme is a self-replicating unit of behavior, like the gene. The meme and the gene are equally selfish. The history of human culture shows us to be as subject to the tyranny of our memes as other species are to the tyranny of genes. But Dawkins ends his discussion with a call for liberation. Our capacity for foresight gives us the power to transcend our memes, just as our culture gave us the power to transcend our genes. We, he says, alone on Earth, can rebel against the tyranny of the selfish replicators.

Dawkins's vision of the human situation as a Promethean struggle against the tyranny of the replicators contains important elements of truth. We are indeed rebels by nature, and his vision explains many aspects of our culture which would otherwise be mysterious. But his account leaves out half of the story. He describes the history of life as the history of replication. Like Eigen, he believes that the beginning of life was a self-replicating molecule. Throughout his history, the replicators are in control. In the beginning, he says, was simplicity. The point of view which I am expounding in this book is precisely opposite. In the beginning, I am saying, was complexity. The essence of life from the beginning was homeostasis based on a complicated web of molecular structures. Life by its very nature is resistant to simplification, whether on the level of single cells or ecological systems or human societies. Life could tolerate a precisely replicating molecular apparatus only by incorporating it into a translation system which allowed the complexity of the molecular web to be expressed in the form of software. After the transfer of complication from hardware to software, life continued to be a complicated interlocking web in which the replicators were only one component.

The replicators were never as firmly in control as Dawkins imagined. In my version the history of life is counterpoint music, a two-part invention with two voices, the voice of the replicators attempting to impose their selfish purposes upon the whole network, and the voice of homeostasis tending to maximize diversity of structure and flexibility of function. The tyranny of the replicators was always mitigated by the more ancient cooperative structure of homeostasis inherent in every organism. The rule of the genes was like the government of the old Hapsburg Empire, "Despotismus gemildert durch Schlamperei" ("Despotism tempered by sloppiness").

One of the most interesting developments in modern genetics is the discovery of "Junk DNA," a substantial component of our cellular inheritance which appears to have no biological function. Junk DNA is nucleic acid which does us no good and no harm, merely taking a free ride in our cells and taking advantage of our efficient replicative apparatus. It is difficult to measure the fraction of our DNA that is functional. Several lines of evidence indicate that as much as half of it may be junk. The prevalence of Junk DNA is a striking example of the sloppiness which life has always embodied in one form or another. It is easy to find in human culture the analogue of Junk DNA. Junk culture is replicated together with memes, just as Junk DNA is replicated together with genes. Junk culture is the rubbish of civilization, television commercials and astrology and jukeboxes and political propaganda. Tolerance of junk is one of life's most essential characteristics. In every sphere of life, whether cultural, economic, ecological or cellular, the systems which survive best are those which are not too fine-tuned to carry a large load of junk. And so, I believe, it must have been at the beginning. I would be surprised if the first living cell were not at least 25 percent junk.

I have been trying to imagine a framework for the origin of life, guided by a personal philosophy which considers the primal characteristics of life to be homeostasis rather than replication, diversity rather than uniformity, the flexibility of the cell rather than the tyranny of the gene, the error tolerance of the whole rather than the precision of the parts. The framework which I have found is an abstract mathematical model, far too simple to be true. But the model incorporates in a crude fashion the qualitative features of life which I consider essential: looseness of structure and tolerance of errors. The model fits into an overall view of life and evolution more relaxed than the traditional view.

The new and looser picture of evolution is strongly supported by recent experimental discoveries in molecular biology. Edward Wilson describes the new picture of the genes in a cell as "a rainforest with many niches occupied by a whole range of elements, all parts of which are in a dynamic state of change." The metaphor of the rain forest, which we applied earlier to the luxuriant growth of particle physics, now reappears in molecular biology. My philosophical bias leads me to believe that Wilson's metaphor describes not only the modern cell but the evolution of life all the way back to the beginning. I hold the creativity of quasi-random complicated structures to be a more important driving force of evolution than the Darwinian competition of replicating monads. But philosophy is nothing but empty words if it is not capable of being tested by experiment. If my remarks have any value, it is only insofar as they suggest new experiments. I leave it now to the experimenters to see whether they can condense some solid facts out of this philosophical hot air.

6

HOW WILL IT ALL END?

In Chapter 3, I said that I wanted to find connections between cosmology and biology, between the architecture of the universe and the existence of its living inhabitants. But then the next two chapters discussed the nature and origin of life without any further mention of cosmology. In separating the origin of life from the origin of the universe, I was following the rule which every student of modern science is taught to follow: local events are to be explained by local causes. The origin of life is assumed to have been a local event, occurring on a particular planet at a particular time. The age of the Earth is known to be less than half of the age of the universe. The origin of life was far removed in time and in physical conditions from the days when the universe was young. Therefore, following the rule of local causation, we seek to explain the origin of life as a consequence of physical and chemical processes here on earth, unaffected by any teleological influences from the remote past or from the remote future.

If a theory of the origin of life is successful, it means that a particular chain of hypothetical local events is found to lead to the emergence of life with reasonable probability. Theories of the origin of the universe are the concern of a separate group of experts in a separate scientific discipline. The separation between the disciplines of biology and cosmology may be philosophically regrettable but it is built into the structure of modern science. All our cosmological theories assume that the early history of the universe was driven by purely physical processes in which foreshadowings of life played no part.

I will not describe in detail what we know about the early universe. One reason I leave the early universe aside is that it seems to have been biologically barren and not directly relevant to my theme. The other reason is that there is an excellent book, *The First Three Minutes,* by Steven Weinberg, which gives an account of the state of our knowledge about the beginning of the universe, written in lucid nontechnical language. Although Weinberg's book was published in 1977, almost everything in it is still true. I have some disagreement with Weinberg's philosophy, but I have no disagreement with his facts.

One of the facts to which Weinberg calls attention is the timidity of theorists. Before 1965, when Arno Penzias and Robert Wilson discovered the microwave background radiation pervading the universe, nobody took theories of the early universe seriously. Even the people who invented theories did not take them seriously. George Gamow, who correctly predicted the existence of the background radiation in 1948, was generally considered to be a popularizer of science rather than a serious scientist. As Weinberg remarks, progress in understanding and observing the universe was delayed for a long time by the timidity of theorists:

> This is often the way it is in physics. Our mistake is not that we take our theories too seriously, but that we do not take them seriously enough. It is always hard to realize that these numbers and equations we play with at our desks have something to do with the real world. Even worse, there often seems to be a general agreement that certain phenomena are just not fit subjects for respectable theoretical and experimental effort. . . . The most important thing accomplished by the discovery of the radiation background

in 1965 was to force all of us to take seriously the idea that there *was* an early universe.

Thanks to Penzias and Wilson, Weinberg and others, the study of the beginning of the universe is now respectable. Professional physicists who investigate the first three minutes or the first microsecond no longer need to feel shy when they talk about their work. But the end of the universe is another matter. I have searched the literature for papers about the end of the universe and found very few. And the striking thing about these papers is that they are written in an apologetic or jocular style, as if the authors were begging us not to take them seriously. The study of the remote future still seems to be as disreputable today as the study of the remote past was thirty years ago. I am particularly indebted to Jamal Islam, a physicist colleague now living in Bangladesh, for an early draft of his 1977 paper which started me thinking about the remote future. Islam and I are hoping to hasten the arrival of the day when eschatology, the study of the end of the universe, will be a respectable scientific discipline and not merely a branch of theology.

Weinberg himself is not immune to the prejudices that I am trying to dispel. At the end of his book about the past history of the universe, he adds a short chapter about the future. He takes 150 pages to describe the first three minutes, and then dismisses the whole of the future in five pages. Without any discussion of technical details, he sums up his view of the future in twelve memorable words: "The more the universe seems comprehensible, the more it also seems pointless."

Weinberg has here, perhaps unintentionally, identified a real problem. It is impossible to calculate in detail the long-range future of the universe without including the effects of life and intelligence. It is impossible to calculate the capabilities of life and intelligence without touching, at least peripherally, philosophical questions. If we are to examine how intelligent life may be able to guide the physical development of the universe for its own purposes, we cannot altogether avoid considering what the values and purposes of intelligent life may be. But as soon as we mention the words "value" and "purpose," we run into one of the most firmly entrenched taboos of twentieth-century science. Hear the voice of Jacques Monod, high priest of scientific rationality, in his book *Chance and Necessity:* "Any mingling of knowledge with values is unlawful, forbidden."

Monod was one of the seminal minds in the flowering of molecular biology in this century. It takes some courage to defy his anathema. But I will defy him and encourage others to do so. The taboo against mixing knowledge with values arose during the nineteenth century out of the great battle between the evolutionary biologists led by Thomas Huxley and the churchmen led by Bishop Wilberforce. Huxley won the battle, but a hundred years later Monod and Weinberg were still fighting Bishop Wilberforce's ghost. Physicists today have no reason to be afraid of Wilberforce's ghost. If our analysis of the long-range future leads us to raise questions related to the ultimate meaning and purpose of life, then let us examine these questions boldly and without embarrassment. If our answers to these questions are naive and preliminary, so much the better for the continued vitality of our science.

I described in Chapter 3 an older example of the timidity of theorists: Isaac Newton, the greatest theorist of his age, suppressed and concealed the writings in which he went outside the bounds of scientific orthodoxy. In spite of his immense achievements and worldly success, he remained at heart a timid soul. He was a secret Unitarian, afraid to risk his tenured position at Cambridge University by expressing in public the unorthodox theology which is to be found in his private papers. And he was a secret cosmologist, afraid to risk the disapproval of his scientific colleagues by expressing in public his private thoughts about the place of life in the universe.

It is fruitless to speculate upon the might-have-beens of history. What might have been the effect on the cultural history of Europe in the eighteenth and nineteenth centuries, if Newton had had the courage to publish his speculative cosmology? Judging from the state of the manuscript, which is a finished copy accompanied by several rough drafts, he had at one time intended to publish it. If he had done so, placing his unparalleled prestige as the supreme intellect of the Age of Enlightenment behind an unashamedly romantic and poetic view of the cosmos, would it have made any difference? Would European culture have avoided that whole disastrous split between the narrow rationalism of the Enlightenment, dominated by Newton's public image, on the one hand, and the excessive irrationalism of the romantic reaction on the other? Could we have avoided the political manifestations of this split in Napoleonic centralism on the one hand and nationalistic frenzy on the other? Could we have avoided the conflict between science and religion, the conflict which soured the intellectual life of the nineteenth century and continues to impoverish Western culture even today? It is unfair to hold Newton responsible for all these later misfortunes. It was not his fault that the universe beyond the Earth turned out to be dead and empty.

Perhaps it is inevitable that every great theorist, exhausted by the supreme effort of creating a new scientific vision, should be afraid of exploring still further into the surrounding darkness. The timidity of theorists is a natural human reaction. After a great discovery, the discoverer would like to settle down and consolidate the new territory. It is natural then to set limits, to draw lines beyond which further speculation should not go. Perhaps we should not be surprised that Darwin, the theorist who dominated the nineteenth century as Newton dominated the eighteenth, was also on many occasions as timid as Newton. Darwin was a less secretive character than Newton, but he was equally hesitant in publishing his ideas. There is a curious parallelism between Darwin's twenty-year delay in publishing his theory of evolution and Newton's twenty-year delay in publishing the *Principia*. And Newton's refusal to publish his cosmological speculations finds a parallel in Darwin's silence concerning the problem of the origin of life.

If we are to understand in general terms the place of life in the universe, we must also understand life's origin. Darwin explicitly excluded the origin of life from the scope of his theory of evolution. In his earliest outline of the theory, drawn up in 1839 or soon thereafter, he wrote, "Extent of my theory—having nothing to do with first origin of life, grow [sic], multiplication, mind (or with any attempt to find out whether descended from one form and what that form was)." He knew that only a massive and meticulous marshaling of evidence could make his theory convincing to his contemporaries; he had no hope of finding observational evidence relevant to the origin of life, and he therefore made no claim that the theory might throw light on the question of origins. Any such claim would expose a weak point in the structure of facts upon which his argument rested. He could, and did, discuss the origin of life privately among his friends. He could not discuss it publicly for fear of handing to his enemies the choice of a battleground which he could not defend with factual evidence. So, like Newton, Darwin kept his more radical thoughts private and presented to the world a scientific edifice of grand design but limited scope. "Hypotheses non fingo" became his motto, too. The effect of Darwin's caution was similar to the effect of Newton's. In both cases, a new science was created with clearly demarcated boundaries, and speculation transcending the boundaries was severely discouraged. The separation between biology and cosmology was reinforced by territorial boundaries on both sides. The question of the origin of life, which should have been a common meeting ground for the practitioners of both disciplines, became a no-man's-land accepted by neither.

We are fortunate to be living at a time when the barriers are breaking down. The collapse of the barriers was celebrated a few years ago at a meeting in Cambridge, where biologists as respectable as Sydney Brenner mingled with cosmologists as respectable as Stephen Hawking. The title of the meeting was "Matter into Life," and the discussions were memorable not for the number of problems solved but for the number of problems unsolved. A huge territory now lies open, ranging from the crystallography of clay at one end to neurophysiology at the other, and over this whole area there is a continuity of unsolved problems ready to sprout into new sciences. There is also a steady stream of pioneers infiltrating the territory from both ends.

At the Cambridge meeting, Eigen and Cairns-Smith talked about the competing theories of the origin of life which I described in Chapter 5. Chemists talked about experiments and mathematicians talked about computer simulations. The new science of the origin of life, having at last achieved respectability, was launched into a period of vigorous growth. Only the cosmologists at the meeting did not say much. The closer the problem of the origin of life comes to being accessible to experimental attack, the less it has to do with cosmology. In the end, the testing of the various theories will be a job for chemists, not for cosmologists.

I came home from the meeting with mixed feelings. On the one hand, the meeting was successful in bringing together the diverse points of view which can contribute to an understanding of life's origins. Biologists and chemists and physicists and mathematicians told us what they knew and learned to understand each other's ways of thinking. On the other hand, the meeting failed to bring together biology and cosmology. Biology and chemistry and physics and mathematics are coming to grips with the problem of the origin of life without any help from cosmology. The meeting confirmed my belief that, if we are interested in bringing biology and cosmology into a single picture of the universe, we must look to the future and not to the past.

Looking to the future, we give up immediately any pretence of being scientifically respectable. From this point on, I make no apology for mixing science with science fiction. Science fiction is, after all, nothing more than the exploration of the future using the tools of science. Unfortunately, science fiction acquired a bad reputation because much of it is poor fiction and even worse science. I will try to keep the fiction plausible and the science accurate.

Progress in understanding the origin of life goes hand in hand with progress in understanding the structure of existing creatures, and progress in biochemistry goes hand in hand with progress in genetic engineering. Newton discarded his youthful dream of a life-filled universe, but the time may soon come when we may be in a position to take matters into our own hands, to make Newton's dream come true by our own efforts. Suppose that we find no radio messages traveling through space, transmitted by extraterrestrial civilizations for our enlightenment. Suppose that we fail to find traces of life anywhere outside our own planet. What then would be the minimum modifications that would have to be imposed upon terrestrial life to enable us to make good nature's lack? Now that genetic engineering is rapidly becoming a practical proposition, it is not absurd to think of redesigning terrestrial creatures so as to make them viable in space or on other celestial bodies. Once we have successfully planted a variety of species in space and provided appropriate mechanisms of dispersal for their spores, we can safely rely on the ancient processes of mutation and natural selection to take care of their subsequent evolution. The mistakes which we shall inevitably make in our initial plantings will in time be rectified as their offspring diversify and spread through the cosmos.

There are three principal obstacles to be overcome in adapting a terrestrial species to life in space. It must learn to live and be happy in zero-g, zero-T, and zero-P, that is to say, zero-gravity, zero-temperature, and zero-pressure. Of these, zero-g is probably the easiest to cope with, although we are still ignorant of the nature of the physiological hazards which it imposes. To deal with zero-T is simple in principle although it may be complicated and awkward in practice. Fur and feathers provide even better insulation in a vacuum than they do in air or in water. Creatures adapted to space must learn to balance the energy generated by their metabolism and by absorption of sunlight or starlight against the energy lost by radiation from their surfaces. Some active control of the radiating surfaces will probably be required. But the low temperature of the environment makes the regulation of an organism's internal temperature easier rather than more difficult. It is easier to keep warm on Pluto than to keep cool on Venus. Once a species has learned to keep warm on Pluto, it will soon be able to keep warm almost anywhere in the universe. The main innovation which the adaptation of life to zero-T requires is that plants as well as animals must become warm-blooded. It seems to be only a curious historical accident that the natural evolution of life on earth produced warm-blooded animals before it produced warm-blooded plants. Only because of this poverty of invention in the plant kingdom do we find the coasts of our Arctic Ocean thinly populated with polar bears rather than thickly populated with living greenhouses. A warm-blooded plant, growing around itself a living greenhouse as effortlessly as a polar bear grows fur or a whale blubber, could spread quickly over the plains of Pluto.

But the most important and most difficult part of the adaptation of life to the cosmos is the adaptation to zero-P. It is only by adapting to zero-pressure, by learning to live in vacuum, that life can be liberated from narrow and malodorous confinement in space capsules and space suits. The move of life from air to vacuum is as fundamental and as liberating as the move which our ancestors made from water to air half a billion years ago. I cannot pretend to have understood the problems of adaptation in detail. I could not define the specifications of a zero-P potato precisely enough for a genetic engineer to sit down and begin working out the DNA sequences to put into its chromosomes. The best I can do is to quote from that old and frustrated engineer Konstantin Tsiolkovsky, who already understood the essentials of the problem of zero-P life when he published his book *Dreams of Earth and Sky* in the Russian provincial town of Kaluga in 1895. The passage which I quote is a dialogue between Tsiolkovsky and the guide who is escorting him around the solar system.

> "These remarkable creatures combine the characteristics of animals and plants and so I call them animal-plants. . . ."
>
> "All right. Don't get angry. Just explain how your creatures avoid getting dried up like mummies."
>
> "That is simple. Their skin is covered with a glassy layer, thin and flexible but absolutely impermeable to gases and liquids and all kinds of particles, so that the creatures are protected from any loss of material. . . . Their bodies have appendages which look like wings and are exposed to the sun, serving as chemical laboratories for the production of energy and food. Such appendages would be an encumbrance in earth's gravity-field, but in the zero-g environment of space they are hardly noticeable even with a surface area of several thousand square meters."
>
> "Stop! How do your animal-plants talk to each other and exchange ideas without any air? . . ."
>
> "They have a much more perfect and natural means of communication. One part of their body carries under the transparent skin an area like a camera obscura, on which moving pictures are continually playing, following the flow of their thoughts and representing them precisely. The pictures are formed by fluids of various colors which flow through a web of fine channels under the skin."

I will leave Tsiolkovsky at this point, dreaming of celestial creatures who communicate with each other in glorious technicolor sign language. This dream is made especially poignant by the fact that Tsiolkovsky himself was cut off from easy communication with his terrestrial fellow creatures by his deafness. For a deaf man, sign language and television are natural channels of communication; radio is not.

Tsiolkovsky might have said of himself what the British Foreign Secretary George Canning said in 1826 when he had broken away from the alliance of reactionary monarchical governments in Europe and recognized the independence of the new Latin American republics: "I look at the Indies and I call in the new world to redress the balance of the old." Tsiolkovsky called into existence the new world of cosmic biology, toward which his invention of interplanetary rocketry was only a stepping stone, to redress the balance between living and non-living in the old world of natural cosmology, to redress the balance that was destroyed when Newton consigned his more speculative manuscripts to the attic. Newton banished life from the celestial spaces. Tsiolkovsky showed us the way to bring it back. Newton was a unifier, Tsiolkovsky a diversifier. If we are striving to reach a balanced view of the universe, we need to listen to them both.

For the remainder of this chapter I shall assume that life is able to solve the technical problems of adapting itself to zero-g, zero-t, and zero-p. I assume that life is capable of making itself at home in every corner of the universe, just as it has made itself at home in every corner of this planet. I wish now to go beyond qualitative statements and begin a quantitative assessment of the potentialities of life in the universe. For a quantitative assessment, even at the rudimentary level which is all I can hope to achieve, some degree of abstraction is required. In order to make quantitative statements which are independent of local accidents, we need to introduce an abstract description of living processes independent of mechanical and chemical details. My argument will be based on a fundamental assumption concerning the nature of life, that life resides in organization rather than in substance. I am assuming that my consciousness is inherent in the way the molecules in my head are organized, not in the substance of the molecules themselves. If this assumption is true, that life is organization rather than substance, then it makes sense to imagine life detached from flesh and blood and embodied in networks of superconducting circuitry or in interstellar dust clouds.

Let me now state my assumptions in a precise and quantitative fashion. I call the assumption that life can be described abstractly the hypothesis of abstraction. The hypothesis of abstraction says that every living creature is characterized by a number Q which is a measure of the complexity of the creature. To measure Q, we do not need to know anything about the internal structure of the creature. Q can be measured by observing from the outside the behavior of the creature and its interaction with its environment. Q is simply the quantity of entropy produced by the creature's metabolism during the time it takes to perform an elementary response to a stimulus. "Entropy" is a jargon word meaning waste heat divided by temperature. By an elementary response, I mean a simple action such as moving when poked or yelling when hurt. When the quantity of entropy is expressed in bits of information, the natural unit of entropy according to the law of equivalence of entropy and information, the quantity Q becomes a pure number.

For a human being, Q turns out to be about 10^{23}, which is a rough measure of the amount of waste heat we have to generate in order to do anything at all. It measures the number of bits of information that must be processed to keep a person alive long enough to say, "Cogito, ergo sum." It is probably not an accident that Q happens also to be roughly equal to the number of large molecules in a human body. But the hypothesis of abstraction says that we can make general statements about life in the abstract, not knowing anything about its molecular structure, using the externally measured Q as a sufficient description of its hidden complexity. Complexity is an extensive quantity; that is to say, the complexity of a group of creatures is the sum of the complexities of the individuals in the group. The complexity that can be achieved by life in a given environment is ultimately proportional to the quantity of matter available for its use.

In addition to the hypothesis of abstraction, I make a much stronger assumption which I call the hypothesis of adaptability. The hypothesis of adaptability says that, given sufficient time, life can adapt itself to any environment whatever. This is certainly an exaggeration of the truth. But it is of interest to examine first the idealized case of perfect adaptability. Practical limitations and qualifications can always be added later. The formal statement of the hypothesis of adaptability is then as follows. Suppose that we have two environments A and B with a sufficient supply of matter and energy but with different environmental conditions of temperature, pressure, chemistry and so on. If a creature of complexity Q can exist in A, then an equivalent creature with the same complexity Q can exist in B. The word "equivalent" means that the creature in B has the same behavior patterns as the creature in A. If one is intelligent, the other is equally intelligent. If one is conscious, the other is also conscious. If one of them can talk to its friends about the content of its consciousness, the other will describe in an equivalent language a subjectively identical consciousness.

The hypotheses of abstraction and adaptability have some similarity to the first and second laws of thermodynamics. They are simple statements of a qualitative nature, and yet they provide a foundation for a quantitative theory of cosmic ecology. I will not burden you with the mathematical details of the theory.

The main theorem of cosmic ecology is that for a life form of fixed complexity adapting itself to a variety of environments, the rate of metabolism of energy varies with the square of the temperature. This fact, that energy consumption goes like temperature squared, has important consequences. It means that cold environments are fundamentally more hospitable to complex forms of life than hot environments. Life is, after all, an ordered form of matter, and low temperature favors order. In the long run, life depends less on an abundant supply of energy than on a good signal-to-noise ratio. The colder the environment, the quieter the background, the more thrifty life can be in its use of energy.

How far life can go depends not only on biological adaptability but also on physical cosmology. If the physical universe is finite in space and time, the possible scope of life is also finite. In this case, as John Wheeler graphically describes it, the universe which begins its existence with a Big Bang ends its existence with a big crunch. The big crunch is a universal collapse into heat-death, with the sky growing hotter and hotter until it finally falls in on us and carries us into a space-time singularity at infinite temperature. Nothing living can survive the big crunch. If our universe is finite, then life will barely have time to spread itself once around the cosmos before the inexorable collapse begins. If that is to be our fate, then life is as the poet Housman described it:

> Here on the level sand
> Between the sea and land,
> What shall I build or write
> Against the fall of night?
> Tell me of runes to grave
> That hold the bursting wave,
> Or bastions to design
> For longer date than mine.

There is a great melancholy in the picture of a finite universe, its life force spent, its days of passion over, counting the hours remaining before it slides into oblivion. What will our last poets sing, whoever they may be, human or alien, as they watch the stars crowding together and streaming faster and faster across the imploding sky? Perhaps in their final moments they will remember the words of our contemporary, Ivor Gurney, echoing down the eons from the springtime of our species:

> The songs I had are withered
>   or vanished clean,
> Yet there are bright tracks
>   Where I have been,
> And there grow flowers
>   For others' delight.
> Think well, O singer,
>   Soon comes night.

If, on the other hand, we live in an open universe, infinite in space and time and continuing to expand into a future without end, then life has to face a prospect of slow freezing rather than quick frying. The universe grows constantly colder as it expands, and the supply of free energy is constantly dwindling. To many people this future of endless ice has seemed even more dismal than the future of cataclysmic fire. But the laws of cosmic ecology put these futures into a very different perspective. If the hypothesis of adaptability is correct, life has a clear preference for ice over fire. In an expanding universe, life can adapt itself as the eons go by, constantly matching its metabolism of energy to the falling temperature of its surroundings. Since we are assuming perfect adaptability, the rate of energy metabolism falls with the square of the temperature. This has the consequence that, in an expanding universe, life of any fixed degree of complexity can survive forever upon a finite store of energy. The pulse of life will beat more slowly as the temperature falls but will never stop.

My theoretical physicist colleagues have recently found serious reasons to believe that all matter may be unstable. According to their latest theoretical models, the Grand Unified models of particle physics, the nuclei of all atoms will disappear into positrons and photons and neutrinos with a lifetime of the order of 10^{33} years. The reality of this universal decay of matter into radiation is now to be put to experimental test. Within a few years we should know for sure whether matter is permanent or transient. If the experiment decides that matter is transient, then life will have to face some severe problems about 10^{33} years from now. By human standards 10^{33} years is a long time, but in the eye of eternity it is but an instant. If the universe is open, the history of life will be only just beginning when the disappearance of matter threatens its existence.

This will be the supreme test of life's adaptability. I do not know whether we can survive without protons. But I do not see any reason even then to declare the situation hopeless. If the assumptions of abstraction and adaptability are correct, the patterns of life and consciousness should be transferable without loss from one medium to another. After the protons are gone, we shall still have electrons and positrons and photons, and immaterial plasma may do as well as flesh and blood as a vehicle for the patterns of our thought. Perhaps the best possible universe is a universe of constant challenges, a universe in which survival is possible but not too easy. If optimism is the philosophy of people who welcome challenges, then we have plenty of reason to be optimists.

One of the great scientists of this century was the crystal-lographer Desmond Bernal, who began looking at the crystal structure of nucleic acids with X-rays long before DNA became fashionable. Bernal was also a gifted writer and philosopher. In the year 1929 he wrote a little book, *The World, the Flesh and the Devil; an Inquiry into the Three Enemies of the Rational Soul.* He explored with extraordinary vision and boldness the possible future of mankind, and the ways in which we may defeat the three traditional enemies of our progress: the World symbolizing material obstacles, famine and flood and poverty; the Flesh symbolizing our bodily infirmities, disease and senility and death; and the Devil symbolizing the dark irrational forces within our own minds, greed and fear and jealousy and madness. His vision went far into the future, and at the end he described what he saw as the ultimate transformation of life in the universe. Bernal anticipated some of the discoveries which biologists and physicists were to make fifty years later. In particular, he foresaw and welcomed the transition of life from material substance into etheriality which, if the Grand Unified models are right, the decay of protons will make inevitable:

> One may picture, then, these beings, nuclearly resident, so to speak, in a relatively small set of mental units, each utilizing the bare minimum of energy, connected together by a complex of etherial intercommunication, and spreading themselves over immense areas and periods of time by means of inert sense organs which, like the field of their active operations, would be, in general, at a great distance from themselves. As the scene of life would be more the cold emptiness of space than the warm, dense atmosphere

of the planets, the advantage of containing no organic material at all, so as to be independent of both of these conditions, would be increasingly felt. . . .

Bit by bit the heritage in the direct line of mankind, the heritage of the original life emerging on the face of the world, would dwindle, and in the end disappear effectively, being preserved perhaps as some curious relic, while the new life which conserves none of the substance and all the spirit of the old would take its place and continue its development. Such a change would be as important as that in which life first appeared on the earth's surface and might be as gradual and imperceptible. Finally, consciousness itself may end or vanish in a humanity that has become completely etherialized, losing the close-knit organism, becoming masses of atoms in space communicating by radiation, and ultimately perhaps resolving itself entirely into light. That may be an end or a beginning, but from here it is out of sight.

That is Bernal writing in 1929. From here it is out of sight. Or, if you find Bernal's description too optimistic for our present mood of doom and gloom, there is the more sardonic but equally bold statement of the biologist J. B. S. Haldane. Haldane was one of the angry young men who fought in the trenches in World War I and came out of that inferno with a bitter laugh. Writing in 1923, six years before Bernal, he summarized his view of the human situation in a little book called *Daedalus, or Science and the Future.* Here is Haldane's verdict on the prospects of life in the universe:

The human intellect is feeble, and there are times when it does not assert the infinity of its claims. But even then—

Though in black jest it bows and nods,
I know it is roaring at the Gods,
Waiting the last Eclipse.

When I published my speculations about the future of the universe in the *Reviews of Modern Physics,* the literary quotations were put in as occasional embellishments to break the monotony of the main argument. The main argument was a chain of mathematical calculations beginning with equation 1 and ending with equation 137. This chapter contains the embellishments without the equations. There is a danger that the argument presented here may seem to be nothing but literary frills without any scientific substance. The mathematical core of the argument was solid enough to be approved by the referees of a professional physics journal. The reader will have to take my word for it that the equations were right. I cannot explain here everything that went into the equations. Instead, I will summarize briefly the results that came out at the end.

I was attempting to find answers to three questions within the framework of an open and indefinitely expanding universe.

(1) Does the universe freeze into a state of permanent physical quiescence as it expands and cools?

(2) Is it possible for life and intelligence to survive forever?

(3) Is it possible to maintain communication and transmit information across the constantly expanding distances between galaxies?

Tentatively, I answer these three questions with a no, a yes, and a maybe. No, the laws of physics do not predict any final quiescence but show us things continuing to happen, physical processes continuing to operate, as far into the future as we can imagine. Yes, life and intelligence are potentially immortal, with resources of knowledge and memory constantly growing as the temperature of the universe decreases and the reserves of free energy dwindle. And maybe, intelligent beings in different parts of the universe can keep alive forever a network of communication, exchanging their ideas and constantly increasing their circle of acquaintances. The third answer is tentative because it assumes that transmitters and receivers of information can always be built with efficiency close to the theoretical ideal. The theoretical ideal of efficiency can be calculated from the laws of physics and information theory. To find out whether real transmitters and receivers could approach the ideal performance, one would need to be an electrical engineer rather than a physicist.

The numerical results of my calculations show that the quantities of energy required for permanent survival and communication are surprisingly modest. For a society with the same complexity as the present human society on Earth, starting from the present time and continuing forever, the total reserve of energy required is about equal to the energy now radiated by the Sun in eight hours. The total energy reserve contained in the Sun would be sufficient to support forever a society with a complexity 10 trillion times greater than our own. The energy required to keep open forever a high-speed intergalactic communication system is only about 1 gigawatt-year per channel, an extremely small quantity of energy by astronomical standards. The energy reserve contained in the Sun would be sufficient to keep open forever as many communication channels as would be needed to keep us talking with every star in the visible part of the universe. These statements are based on rough numerical estimates which may easily be wrong by a factor of ten or a hundred. Nevertheless, they give strong support to an optimistic view of the potentialities of life. They imply that the world of physics and astronomy is inexhaustible. No matter how far we go into the future, there will always be new things happening, new information coming in, new worlds to explore, a constantly expanding domain of life, consciousness and memory.

When I write in this style, I am mixing knowledge with values, disobeying Monod's prohibition. But I am in good company. Before the days of Darwin and Huxley and Bishop Wilberforce, in the eighteenth century, scientists were not subject to any taboo against mixing science and values. When Thomas Wright, the discoverer of galaxies, announced his discovery in 1750, he was not afraid to use a theological argument to support an astronomical theory.

> Since as the Creation is, so is the Creator also magnified, we may conclude in consequence of an infinity, and an infinite all-active power, that as the visible creation is supposed to be full of siderial systems and planetary worlds, so on, in like similar manner, the endless immensity is an unlimited plenum of creations not unlike the known. . . . That this in all probability may be the real case, is in some degree made evident by the many cloudy spots, just perceivable by us, as far without our starry Regions, in which tho' visibly luminous spaces, no one star or particular constituent body can possibly be distinguished; those in all likelyhood may be external creation, bordering upon the known one, too remote for even our telescopes to reach.

Thirty-five years later, Wright's speculations were confirmed by William Herschel's precise observations. Wright also computed the number of habitable worlds in our galaxy: "In all together then we may safely reckon 170,000,000, and yet be much within compass, exclusive of the comets which I judge to be by far the most numerous part of the creation."

Wright's statement about the comets may also be correct, although he does not tell us how he estimated their number. For him the existence of so many habitable worlds was not just a scientific hypothesis but a cause for moral reflection:

> In this great celestial creation, the catastrophy of a world, such as ours, or even the total dissolution of a system of worlds, may possibly be no more to the great Author of Nature, than the most common accident in life with us, and in all probability such final and general Doomsdays may be as frequent there, as even Birthdays or mortality with us upon the earth. This idea has something so chearful in it, that I own I can never look upon the stars without wondering why the whole world does not become astronomers; and that men endowed with sense and reason should neglect a science they are naturally so much interested in, and so capable of inlarging their understanding, as next to a dem-

onstration must convince them of their immortality, and reconcile them to all those little difficulties incident to human nature, without the least anxiety.

All this the vast apparent provision in the starry mansions seem to promise: What ought we then not to do, to preserve our natural birthright to it and to merit such inheritance, which alas we think created all to gratify alone a race of vain-glorious gigantic beings, while they are confined to this world, chained like so many atoms to a grain of sand.

There speaks the eighteenth century. But Steven Weinberg says, "The more the universe seems comprehensible, the more it also seems pointless." If Weinberg is speaking for the twentieth century, then I prefer the eighteenth.

When Lord Gifford established these lectures, he directed that they be concerned with Natural Theology. The passages which I quoted from Thomas Wright fit splendidly into the framework of Lord Gifford's directive. I do not know precisely what Lord Gifford had in mind when he spoke of Natural Theology, but I imagine he meant to encourage that combination of scientific learning with moral eloquence which we find in Wright. Thomas Wright, if he were alive today, would have been the ideal Gifford Lecturer.

But I do not wish to leave you with the impression that Wright's theology is a quaint relic of the eighteenth century, irrelevant to our modern science, or that Weinberg's nihilistic voice speaks for all twentieth-century scientists. The universe that I have explored in a preliminary way in this book is very different from the universe which Weinberg envisaged when he called it pointless. I have found a universe growing without limit in richness and complexity, a universe of life surviving forever and making itself known to its neighbors across unimaginable gulfs of space and time. Whether the details of my calculations turn out to be correct or not, there are good scientific reasons for taking seriously the possibility that life and intelligence can succeed in molding this universe of ours to their own purposes. Twentieth-century science, when it looks to the future, provides a solid foundation for a philosophy of hope. A rational soul, knowing what we now know about the universe, has no reason to dismiss as fantasy the optimistic visions of Bernal and of Thomas Wright.

To conclude this chapter I venture a little further than Thomas Wright in the direction of theological speculation. To me the most astounding fact in the universe, even more astounding than the flight of the Monarch butterfly, is the power of mind which drives my fingers as I write these words. Somehow, by natural processes still totally mysterious, a million butterfly brains working together in a human skull have the power to dream, to calculate, to see and to hear, to speak and to listen, to translate thoughts and feelings into marks on paper which other brains can interpret. Mind, through the long course of biological evolution, has established itself as a moving force in our little corner of the universe. Here on this small planet, mind has infiltrated matter and has taken control.

It appears to me that the tendency of mind to infiltrate and control matter is a law of nature. Individual minds die and individual planets may be destroyed. But, as Thomas Wright said, "The catastrophy of a world, such as ours, or even the total dissolution of a system of worlds, may possibly be no more to the great Author of Nature, than the most common accident of life with us." The infiltration of mind into the universe will not be permanently halted by any catastrophe or by any barrier that I can imagine. If our species does not choose to lead the way, others will do so, or may have already done so. If our species is extinguished, others will be wiser or luckier. Mind is patient. Mind has waited for 3 billion years on this planet before composing its first string quartet. It may have to wait for another 3 billion years before it spreads all over the galaxy. I do not expect that it will have to wait so long. But if necessary, it will wait. The universe is like a fertile soil spread out all around us, ready for the seeds of mind to sprout and grow. Ultimately, late or soon, mind will come into its heritage.

What will mind choose to do when it informs and controls

the universe? That is a question which we cannot hope to answer. When mind has expanded its physical reach and its biological organization by many powers of ten beyond the human scale, we can no more expect to understand its thoughts and dreams than a Monarch butterfly can understand ours. Mind can answer our question only as God answered Job out of the whirlwind: "Who is this that darkeneth counsel by words without knowledge?" In contemplating the future of mind in the universe, we have exhausted the resources of our puny human science. This is the point at which science ends and theology begins.

Like the majority of scientists in this century, I have not concerned myself seriously with theology. Theology is a foreign language which we have not taken the trouble to learn. My personal theology is the theology of an amateur. But I did once have some help from a professional theologian in formulating my ideas in an intellectually coherent fashion. I happened to meet Charles Hartshorne at a meeting in Minnesota and we had a serious conversation. After we had talked for a while he informed me that my theological standpoint is Socinian. Socinus was an Italian heretic who lived in the sixteenth century. If I remember correctly what Hartshorne said, the main tenet of the Socinian heresy is that God is neither omniscient nor omnipotent. He learns and grows as the universe unfolds. I do not pretend to understand the theological subtleties to which this doctrine leads if one analyzes it in detail. I merely find it congenial, and consistent with scientific common sense. I do not make any clear distinction between mind and God. God is what mind becomes when it has passed beyond the scale of our comprehension. God may be considered to be either a world-soul or a collection of world-souls. We are the chief inlets of God on this planet at the present stage of his development. We may later grow with him as he grows, or we may be left behind. As Bernal said: "That may be an end or a beginning, but from here it is out of sight." If we are left behind, it is an end. If we keep growing, it is a beginning.

The great virtue of my version of the Socinian theology is that it leaves room at the top for diversity. Just as the greatness of the creation lies in its diversity, so does also the greatness of the creator. Many world-souls are better than one. When mind grows to fill the universe, it comes as a diversifier as well as a unifier.

Another theologian, with whom I have a more distant acquaintance, is St. Paul. St. Paul had some good things to say about diversity. "Now there are diversities of gifts, but the same spirit. And there are differences of administrations, but the same Lord. And there are diversities of operations, but it is the same God which worketh all in all." That passage from First Corinthians would make a good text for my sermon if I were preaching in church. But I am not preaching a Christian sermon. I am describing the universe as I encounter it in my life as a scientist and as a politically engaged citizen. I should not pretend to agree with St. Paul when in fact I find his point of view alien. For St. Paul, the diversity of the creation is less important than the unity of the creator. For me, it is the other way round. I do not know or particularly care whether the same God is working all in all. I care deeply for the diversity of his working.

Once upon a time, about thirty-five years ago, I went to see a play called *Spring 1600* by Emlyn Williams. I have never seen it again and I do not know whether it is a good play, but it left a deep impression. The hero is the actor Richard Burbage, who is presented as he probably was in real life, a rough and impetuous fellow with a heart of gold. He has a wife who looks as if she may have given Shakespeare the idea for the character of Mistress Quickly in *Henry IV*. There is a lot of noise offstage, thumping and hammering. Across the street from the Burbages' home in South London the Globe Theater is being built, the theater in which Richard is destined, during the next ten years, to be playing for the first time the roles of Hamlet, Othello, King Lear, Macbeth. But in Spring 1600 this is all far in the future. They have just had a great success with *Henry V* and are getting things ready for *Twelfth Night*. Will

Shakespeare himself never appears on stage. He is busy taking care of the new theater and negotiating with the people at court for the Queen's patronage. Through the Burbages' home pours a constant stream of members of the company, a turbulent and undisciplined lot, held together only by a shared respect for Mistress Burbage's common sense and Will Shakespeare's genius.

The last act is the opening night of the new theater with the first performance of *Twelfth Night.* We do not see the performance. We see only the desperate last-minute improvisations which barely succeed in getting the show together in time for the opening. The show is, of course, a huge success. The Queen herself is there. In the final scene, after the play is finished and the Queen and the crowds have gone home, Richard goes alone to relax and meditate in the dark and silent theater. After a while he comes back onto our stage. "Did you hear anything?" says his wife. "Yes," says Richard, "I heard whispers of immortality."

That is perhaps too sentimental an ending for our modern taste. But I still like it. We know that when Will Shakespeare died in 1616 he left twenty-six shillings and eightpence to Richard to buy a ring, as a token of his affection. It is not absurd to imagine Richard standing in that empty theater in Spring 1600 and hearing those whispers. So I will use the same ending for my cosmology. We know very little yet about the potentialities and the destiny of life in the universe. In speculating about these matters we follow a great tradition. We are in the same company with Bernal and Newton, Tsiolkovsky and Thomas Wright. Letting our imagination wander among the stars, we too may hear whispers of immortality.

PART TWO

PEOPLE AND MACHINES

As I look round this room, at the bed, at the counter-
pane, at the books and chairs and the little bottles, and
think that machines made them, I am glad. I am very
glad of the bedstead, of the white enameled iron with
brass rail. As it stands, I rejoice over its essential simplicity. I would not wish it different. Its lines are
straight and parallel, or at right angles, giving a sense
of static motionlessness. Only that which is necessary
is there, whittled down to the minimum. There is
nothing to hurt me or to hinder me; my wish for
something to serve my purpose is perfectly fulfilled.
. . . Wherefore I do honour to the machine and to its
inventor.

—D. H. Lawrence, *"Study of Thomas Hardy,"*
1914.

7

ROOTS

The title "Infinite in All Directions" has two meanings. In Part 1 it meant the unbounded potentialities of the universe as it becomes aware of itself through the action of life and intelligence. In Part 2 it will mean the unbounded responsibilities of mankind as custodian of life upon a small planet. Part 1 was concerned with diversity as a fact, with the diversity of the universe as we observe and attempt to understand it scientifically. Part 2 will be concerned with diversity as a goal, with the diversity of human needs and desires which our technology and our political arrangements are attempting to satisfy. Part 1 looked at life in the universe as a phenomenon for us to explain. Part 2 will look at life in the universe as a heritage for us to cherish and as a destiny for us to earn.

Part 2 is focused upon technology as a force for good and evil in human affairs. The first four chapters deal with peaceful uses of technology, with technology as a tool of economic progress and of scientific exploration. The next five deal with weapons, especially nuclear weapons and the threat which they pose to our existence. Among other things I will discuss three topics which have recently become fashionable: nuclear winter, the political future of Germany, and the Reagan "Star Wars" initiative. In the last two chapters I shall attempt to fit our involvement with technology into a broader vision of the future of mankind.

My vision of our future is a personal one. I took advantage of my position as Gifford Lecturer to tell mankind how I think it ought to behave. But why should I expect mankind to pay attention to my exhortations? Why should any credence be given to the opinions of a mathematical physicist straying outside his field of expertise into problems of ethics and cultural history? Why should anybody take my visions seriously? To answer these questions, Part 2 begins with an examination of intellectual roots. My visions and opinions are not merely personal. I did not invent them. If they deserve respect, it is because they arose out of a cultural heritage that existed long before I was born. They have roots going back into the history which we inherited from our ancestors.

Two hundred years is a short time in the memory of the human species. We are still close kin to Thomas Wright and Benjamin Franklin, to that splendid generation of scholars and statesmen who constituted the Age of Enlightenment and founded the United States of America. Alexander Haley, with his book *Roots,* set us a good example by exploring his personal roots in the eighteenth century. He found and brought to life his great-great-great-great-grandfather Kunta Kinte, who was transported to America in shame and ignominy, and experienced in his flesh the dark cruelties which underlay the polish of eighteenth-century enlightenment. Alex Haley traced the unbroken thread of pride and toughness which brought Kunta Kinte's descendants up from slavery and finally into the mainstream of American life. Stimulated by Alex Haley's example, I also began to look into my roots. I was brought up in England as a pure-bred, middle-class Englishman, as deeply rooted in England as Haley in America. When I came to America as a young man, I imagined that I was the first member of my family who ever settled there. I believed that I was coming to live among strangers. Recently I discovered that I was mistaken. I found out that my great-great-great-grandfather Thomas Oliver was a citizen of Boston. He graduated from Harvard in 1753 while Kunta Kinte was still living in freedom in his native village in Africa. A few years after Kunta Kinte arrived at Annapolis in shame and ignominy, Thomas Oliver left Boston in almost equal shame and ignominy.

Thomas Oliver had the misfortune to be a Tory. He was proscribed and his estate was confiscated by the Commonwealth of Massachusetts. He fled as a refugee to England. He was, according to his biographer, a man of gentle and retiring disposition. He was dragged into politics through a bureaucratic error of the colonial administration in London. Some clerk in London submitted the wrong name to the King, and Thomas Oliver was appointed Lieutenant-Governor of Massachusetts. The administration had intended to give the job to his cousin Peter Oliver, but somehow the names of the two Olivers became mixed up. Such things happen in the best of times. And 1774 was not the best of times. Peter Oliver was the man who ought to have been sitting in the hot seat. He was also a Harvard graduate, he was disciplined as a student for stealing a turkey and a goose, and he later rose to be Chief Justice of Massachusetts. So my great-great-great-grandfather Thomas was made to suffer for Peter's sins. But Thomas, like Kunta Kinte, was a survivor. He knew how to adapt himself to adversity. Like Kunta Kinte, he found a wife among his companions in exile. He married a young lady, Harriet Freeman, whose name I bear. And after six generations he came back to Boston, walking the streets of his old city in my shoes, just as Kunta Kinte returned to his tribe on the banks of the Gambia River in the shoes of Alex Haley. In the name of Thomas Oliver I forgave the Commonwealth of Massachusetts for the injustice that was done to him 210 years ago. And I hope the Commonwealth of Massachusetts has forgiven him for accepting the King's commission at a time when less honorable men kept their heads down and stayed out of trouble.

The subject of this chapter is not biological roots but intellectual roots. My intellectual lineage goes back beyond Thomas Oliver to the roots of America. The New World of

America and the New World of the cosmos beyond the Earth have always been linked in the imagination of mankind. I am not only the biological descendant of Thomas Oliver but also the intellectual descendant of Richard Hakluyt and Jules Verne. Richard Hakluyt was the intellectual father of the English settlement of America. Jules Verne was the intellectual father of the American move into space. Neither Hakluyt nor Verne lived long enough to see his dreams come true. But both of them will be remembered long after those who fulfilled their dreams are forgotten.

Richard Hakluyt was the prototype of the academic promoter of exploration. In the sixteenth century the business of exploration was in the hands of two groups of people, the Adventurers, who were rich and invested their money in expeditions, and the Planters, who were poor and invested their lives. Hakluyt was neither an Adventurer nor a Planter. He held a comfortable academic job at Oxford University and encouraged the Adventurers and the Planters by writing books about their exploits. I am told by my Jewish friends that they define a Zionist to be a Jew who asks a second Jew to give money to pay for a third Jew to go to Israel. If Hakluyt had been a Jew, he would have been a good Zionist. He applauded, without sharing, the courage and self-sacrifice of his compatriots. He also acted as a private consultant to Queen Elizabeth and wrote a confidential report with the title "Particular Discourse on the Western Planting," in which he urged the Queen to large expenditures of public funds for the support of American settlements. Acting in a tradition followed by other governmental authorities in succeeding centuries, the Queen rewarded Richard Hakluyt handsomely for his advice and ignored his recommendations. The "Particular Discourse" disappeared into the Royal Archives and was published by the Maine Historical Society 293 years later. Meanwhile, Richard Hakluyt had found other patrons and had addressed his ideas successfully to a wider audience.

High on the list of Hakluyt's priorities, as on the list of priorities of later academic consultants, was the financial support of higher education. In the dedication of Hakluyt's great work *The Principal Navigations, Voyages, Traffiques and Discoveries of the English Nation* to the Lord High Admiral of England, we find the following sentences:

> Since your Lordship is not ignorant, that ships are to little purpose without skillful seamen, and since seamen are not bred up to perfection of skill in much less time than in the time of two apprenticeships, and since no kind of men of any profession in the commonwealth pass their years in so great and continual hazard of life, and since of so many so few grow to gray hairs; how needful it is, that by way of Lectures and such like instructions, these ought to have a better education, than hitherto they have had, all wise men may easily judge. . . . And that it may appear that this is no vain fancy nor device of mine, it may please your Lordship to understand that the late Emperor Charles the Fifth, considering the rawness of his seamen, and the manifold shipwracks which they sustained in passing and repassing between Spain and the West Indies, with an high reach and great foresight established not only a Pilot Major for the examination of such as sought to take charge of ships in that voyage, but also founded a notable Lecture of the Art of Navigation, which is read to this day in the Contractation House at Seville. The readers of which lecture have not only carefully taught and instructed the Spanish Mariners by word of mouth, but also have published sundry exact and worthy treatises concerning Marine causes, for the direction and encouragement of posterity.

This dedication was written in the same year as young William Shakespeare's patriotic drama *Henry V* and speaks to the same public. It is not difficult to find echoes of Hakluyt's words in reports of committees arguing for increased support of technical education, in England and in America, through the centuries all the way down to our own time.

Hakluyt believed not only in education but in action. He had a grand vision of an English-speaking North America which would rival and ultimately surpass the wealth and power of Spain. "For," he wrote, "not to meddle with the state of Ireland nor that of Guiana, there is under our noses the great and ample country of Virginia, the Inland whereof is found of late to be so sweet and wholesome a climate, so apt and capable of all commodities which Italy, Spain and France can afford, that the Spaniards themselves in their own writings acknowledge the Inland to be a better and richer country than Mexico and New Spain itself." It took our ancestors only two hundred years to prove him right.

I consider myself an intellectual descendant of Hakluyt because I stand in the same relation to the universe at large as he stood to America. I am, like him, an academic promoter of exploration, writing books in the comfort of my home in Princeton, proclaiming the glorious future that awaits us when life shall move out from its home on Earth to explore and embellish the universe. Like him, I earn substantial fees for giving advice to a government which never does what I recommend. Like him, I look beyond the government to the general public for my audience. Like him, I appeal to the patriotic spirit of my countrymen as well as to the enlightenment of humanity at large. Just as Hakluyt at the age of sixteen sat spellbound in front of the maps and books of cosmography collected by his uncle Richard Hakluyt the elder, so I sat spellbound at the Jet Propulsion Laboratory in California in front of the pictures of the moons of Uranus sent home by our spacecraft Voyager.

Marjorie Hope Nicolson has collected in a book with the title *Voyages to the Moon* the literary antecedents of our new age of cosmic exploration. She begins, more or less, with Johannes Kepler's dream of life in the Moon, and ends, more or less, with Jules Verne's French-American expedition to the Moon three hundred years later. She does not mention Konstantin Tsiolkovsky, the true heir of Richard Hakluyt in our century, the man who fired the imagination of Russia with dreams of interplanetary colonies and directly inspired the ambitious space programs of the Soviet Union. She is writing within the Western literary tradition, and in our Western culture it is Jules Verne who came closest to the role of Hakluyt. Verne made a profoundly original contribution to the literature of space travel by establishing the tradition that the heroes of space exploration should be comic rather than tragic. Verne was incapable of pretentious solemnity. His characters are jokers rather than stuffed shirts. His people are funny, and as a result they make space exploration credible. My life was changed when I read Jules Verne at the age of eight. I may not have understood the technical problems of designing a lunar projectile, but I understood the human problems well enough. I understood that these crazy Americans held the future in their hands and that the smart thing for me to do would be to join them. So it was Jules Verne who set me on the road to America and from there onward to Uranus.

Another intellectual ancestor of America comes between Richard Hakluyt and Jules Verne. Our third ancestor is the poet and painter William Blake. An important intellectual root of America is to be found in the writings of Blake. In the 1770s he was an angry young man in London, admiring the American Revolution from afar. At one time he thought seriously of moving to America. Although he never traveled farther than a hundred miles from London, America filled his dreams. Here are some lines which he wrote in a private notebook, not for publication, at a time when the open expression of seditious thoughts was severely punished:

> Why should I care for the men of Thames,
> Or the cheating waves of chartered streams,
> Or shrink at the little blasts of fear
> That the hireling blows into my ear?
> Though born on the cheating banks of Thames,
> Though his waters bathed my infant limbs,
> The Ohio shall wash his stains from me:
> I was born a slave, but I go to be free.

It seems odd today to think of the Ohio River as an image of purity. It seems odder still to think of William Blake settling down as a citizen of Pittsburgh. But Pittsburgh was in Blake's time, and remained until the middle of the nineteenth century, a place of exceptional beauty and exceptional civic virtue. Some would say that Pittsburgh remains today a place of exceptional beauty and exceptional civic virtue. Blake might well have felt himself to be less of an alien in Pittsburgh than in London. If he had settled in Pittsburgh, he would not have felt constrained to hide his visions of America in obscure and allegorical language. He might have become an American national poet, a Walt Whitman of the eighteenth century. His style and cadence seem frequently to be anticipating Whitman.

His one major poem about America was published in 1793. It is called "America, a Prophecy." Blake explained clearly what he meant by the word "Prophecy." He wrote: "Prophets, in the modern sense, have never existed. Jonah was no prophet in the modern sense, for his prophecy of Nineveh failed. Every honest man is a Prophet; he utters his opinion both of private and public matters. Thus: if you go on so, the result is so. He never says, such a thing shall happen let you do what you will. A Prophet is a Seer, not an Arbitrary Dictator."

In his poem, Blake describes America with images which we can still recognize after two hundred years. The political message is muffled, but the passion of the American Revolution rings loud and clear.

> Washington spoke: "Friends of America! look over the
>   Atlantic sea;
> A bended bow is lifted in heaven, and a heavy iron
>   chain
> Descends, link by link, from Albion's cliffs across the
>   sea, to bind
> Brothers and sons of America till our faces pale and
>   yellow,
> Heads depressed, voices weak, eyes downcast, hands
>   work-bruised,

Feet bleeding on the sultry sands, and the furrows of
    the whip
Descend to generations that in future times forget."

The citizens of New York close their books and lock
    their chests;
The mariners of Boston drop their anchors and unlade;
The scribe of Pennsylvania casts his pen upon the earth;
The builder of Virginia throws his hammer down in
    fear.

I know thee, I have found thee, and I will not let thee
    go.
Thou art the image of God who dwells in darkness of
    Africa,
And thou art fallen to give me life in regions of dark
    death.
On my American plains I feel the struggling afflictions
Endured by roots that writhe their arms into the nether
    deep.
I see a Serpent in Canada who courts me to his love,
In Mexico an Eagle, and a Lion in Peru;
I see a Whale in the South Sea, drinking my soul away.

These are just a few fragments which I have picked out of
Blake's poem. For Blake in the eighteenth century, as for
Hakluyt in the sixteenth, the South Sea means the Pacific
Ocean. Blake was not, like Hakluyt, an academic functionary
with a secure income. Hakluyt called himself a preacher.
Blake called himself a prophet. For the launching of any great
enterprise, such as the settlement of America or the expansion
of life into the universe, we need both preachers and prophets.
Preachers like Hakluyt often receive honor in their lifetimes
and die rich. Prophets like Blake usually die poor.

    Since I have had to pay college tuition for five daughters,
I have chosen to play the role of preacher. I preach to all who
will listen the gospel of manifest destiny. The destiny which
I am preaching is not the expansion of a single nation or of a
single species, but the spreading out of life in all its multifari-

ous forms from its confinement on the surface of our small planet to the freedom of a boundless universe. This unimaginably great and diverse universe, in which we occupy one fragile bubble of air, is not destined to remain forever silent. It will one day be buzzing with the murmur of innumerable bees, rustling with the flurry of feathered wings, throbbing with the patter of little human feet. The expansion of life, moving out from Earth into its inheritance, is an even greater theme than the expansion of England across the Atlantic. As Hakluyt wrote that there is under our noses the great and ample country of Virginia, I am saying that there is under our noses the territory of nine planets, forty moons, ten thousand asteroids and a trillion comets.

But that is enough of my preaching. We need now a new prophet to put some passion into these preachings. Perhaps we shall have to wait another two hundred years before we can translate the vision of William Blake into a wider cosmography:

I see a Serpent on Iapetus who courts me to his love;
On Ganymede an Eagle, and a Lion on Miranda;
I see a Whale in the Oort Cloud, drinking my soul away.

The greening of the universe will always remain, like the vision of the American Revolution which Blake saw with his prophet's eye, an unfinished story.

8

QUICK IS BEAUTIFUL

The technologies which have had the most profound effects on human life are usually simple. A good example of a simple technology with profound historical consequences is hay. Nobody knows who invented hay, the idea of cutting grass in the autumn and storing it in large enough quantities to keep horses and cows alive through the winter. All we know is that the technology of hay was unknown to the Roman Empire but was known to every village of medieval Europe. Like many other crucially important technologies, hay emerged anonymously during the so-called Dark Ages. According to the Hay Theory of History, the invention of hay was the decisive event which moved the center of gravity of urban civilization from the Mediterranean basin to Northern and Western Europe. The Roman Empire did not need hay because in a Mediterranean climate the grass grows well enough in winter for animals to graze. North of the Alps, great cities dependent on horses and oxen for motive power could not exist without hay. So it was hay that allowed populations to grow and civilizations to flourish among the forests of Northern Europe. Hay moved the greatness of Rome to Paris and London, and later to Berlin and Moscow and New York.

Another technology with far-reaching effects on human society is knitting. Knitting emerged later than hay but just as anonymously. The historical importance of knitting is explained in an article by Lynn White in the *American Historical Review* of February 1974. The title of the article is "Technology Assessment from the Stance of a Medieval Historian." The first unequivocal evidence of knitting technology is on an altarpiece painted in the last decade of the fourteenth century, now in the Hamburg Kunsthalle. It shows the Virgin Mary knitting a shirt on four needles for the Christ Child. White collects evidence indicating that the invention of knitting made it possible for the first time to keep small children tolerably warm through the Northern winter, that the result of keeping children warm was a substantial decrease in infant mortality, that the decrease in mortality allowed parents to become emotionally more involved with their children, and that the increasing attachment of parents to children led to the appearance of the modern child-centered family. The chain of evidence linking the knitting needle with the playroom and the child psychiatrist is circumstantial but plausible. As White says at the conclusion of his analysis: "Late medieval mothers and grandmothers with clacking needles undoubtedly assessed knitting correctly as regards infant comfort and health, but that in the long run a new notion of relationships within the family would thereby be encouraged could scarcely have been foreseen."

Another technology which White retrospectively assesses is the spinning wheel. The spinning wheel was a Chinese invention. The earliest documentary evidence of its existence is a Chinese painting dating from approximately A.D. 1035; it appears to have reached Europe during the thirteenth century. The spinning wheel led to a rapid expansion of European textile manufacture and to a concomitant growth of commerce. The growth was especially rapid in the linen trade. Falling prices led to an immense increase in the use of linen shirts, sheets, towels and of starched and folded linen coifs decking the heads of fashionable ladies. These were the direct

consequences of the new technology. But the indirect consequences were of even greater importance. Cheap linen meant
an accumulation of linen rags, and the availability of linen rags
meant that paper became cheaper than parchment. By the end
of the thirteenth century, the great majority of manuscripts
were written on paper. There was more paper than the scribes
of Europe could cover with their handwriting. The opportunity was open for an enterprising book publisher in Mainz to
do away with the scribes and use machinery to put words on
paper. In this way the invention of the spinning wheel opened
the way for the invention of the printing press.

All these new technologies—printing, spinning, knitting,
and haymaking—have become a permanent part of the fabric
of modern life. There is no going back to the old ways. The
voices of the victims displaced by the new technologies, the
scribes displaced by Gutenberg, the old-fashioned hand spin-
ners displaced by the spinning wheel, the forest people displaced by hay, have long been silent. We cannot measure even
in retrospect the human costs and benefits of a technological
revolution. We do not possess a utilitarian calculus by which
to weigh the happiness and unhappiness of the people who
were involved in these case histories. Technology assessment
is still an art rather than a science. As Lynn White sums up the
lessons learned from his examples: "Technology assessment,
if it is not to be dangerously misleading, must be based as
much, if not more, on careful discussion of the imponderables
in a total situation as upon the measurable elements."

In this chapter I shall be attempting an assessment of contemporary technologies, within the limits delineated by Lynn
White. The function of technology assessment is not to measure but to warn. We cannot predict quantitatively the value
or the cost of a new technology. What we can do is to look
ahead and identify snags and traps. With luck, we may see the
traps far enough ahead so that we can avoid walking into them.

Great inventions like hay and printing, whatever their
immediate social costs may be, result in a permanent expansion of our horizons, a lasting acquisition of new territory for

human bodies and minds to cultivate. One of my heroes is Michael Pupin, a great inventor who came to America in modern times as a penniless immigrant after spending his youth in a Serbian village. Pupin became rich and famous by inventing a practical transmission line for delivering electric power over long distances. I happen to be a pure scientist who has reaped enormous benefits, intellectually and materially, from the social climate created in America by the generation of inventors to which Pupin belonged. I detest and abhor the academic snobbery which places pure scientists on a higher cultural level than inventors. High culture and human understanding are sometimes to be found among pure scientists, but they are to be found just as frequently among inventors. Pupin's achievements grew out of his intense idealism and his deep faith in the perfectibility of man and in the ennobling influence of science. He wrote an autobiography with the title *From Immigrant to Inventor,* which is the testament of a unique spirit. He enjoyed with equal gusto the peace of his native village, the human comedy of nineteenth-century America, and the impersonal beauty of Maxwell's equations. Here is a passage from a preface which I wrote for his autobiography when it was reprinted:

> Pupin believed with passionate intensity that the primary aim of science is the pure understanding of nature, and that useful applications must be considered of secondary importance. The prestige and influence which he derived from his inventions he used in an unceasing campaign to improve the standing of fundamental science in America. In this way the paradoxical situation arose, that it was Pupin the practical inventor who did more than any other man of his time to convince the American public that great scientific discoveries are more important than inventions. Pupin's triumph has been so complete that now, twenty-five years after his death, pure scientists have more prestige, more influence and more financial support than he would have imagined possible. Perhaps, in this apotheosis of the fundamental researcher,

some injustice has been done to the class of inventors to which Pupin himself belonged. We have reached the point where a first-rate inventor is rarer than a first-rate scientist. Inventors are no longer welcomed in most university departments, and even in industrial laboratories pure research is becoming more and more the fashionable thing to do. Perhaps the time will soon come when a group of pure scientists will be compelled to organize a campaign to prevent the permanent extinction of the inventor.

These words were written twenty-five years ago. Perhaps we are seeing in recent years some signs of an increasing public awareness of the importance of invention. Perhaps the scientific community is becoming aware that its survival depends in the long run upon maintaining a healthy balance between pure science and invention. Perhaps the heroic age of invention may be returning. But we need first to find answers to two questions. Why have our efforts to apply science fruitfully to human needs in recent decades been so conspicuously unsuccessful? And what can we now do to make things go better?

We often hear it said that there is a similarity between the human condition of the nuclear physicists forty years ago and that of the biochemists and biologists today. There are many obvious differences between nuclear physics and microbiology, but the analogies are real. We are now entering a period of intensified biological research and are considering a variety of possible applications of genetic engineering. Perhaps the historical experience of the physicists may have left us with some practical wisdom which might enable the genetic engineering industry to avoid the mistakes that have brought the nuclear power industry into such serious trouble. I will try to put on record here what little wisdom I have collected in forty years as a part-time applied physicist. Let me come at once to the two questions I am trying to answer. What did the nuclear physicists do wrong? And what can the genetic engineers learn from our misfortunes? I will tell a few stories about things that I have seen happen during my life as a physicist, and the reader can then judge whether these stories have any relevance to the problems of genetic engineering.

The first story concerns a company called General Atomic which runs a laboratory in La Jolla, California, and manufactures nuclear reactors. The company began as a division of the General Dynamics Corporation in the year 1956. In the summer of that year the company brought together a group of consultants, some expert and some non-expert in the details of reactor engineering, and paid us to sit and think for three months. I was one of the non-expert consultants. The company was then brand new; it had no laboratories, no production facilities, and no products. The consultants could do nothing except think and talk and scribble on blackboards. The company promised to pay one dollar to the inventor for the patent rights to any reactor which we might invent. I collected my dollar, and so did several other people in the group. In return for this substantial outlay, the company ended the summer with preliminary designs for three new types of reactor with some promise of commercial profitability. One of these designs was chosen for immediate development and went into production with the name TRIGA, standing for Training, Research and Isotope production, General Atomic. The first TRIGA was built, tested, licensed, and sold within less than three years from the day the consultants assembled in 1956. The company is still producing the TRIGA and still selling it at a profit. The TRIGA is not a power reactor; it is mostly used to produce isotopes for medical research and diagnosis, not to produce electricity.

As a follow-on to the TRIGA, General Atomic decided to develop and market a big power reactor called HTGR, High-Temperature Gas-Cooled Reactor. The HTGR is theoretically a great reactor. Its high temperature gives it an advantage in thermodynamic efficiency over water-cooled reactors, and its big heat capacity gives it an advantage in safety. It is inherently much less vulnerable to mishandling than the light-water reactors which have monopolized nuclear power production in the

United States. And since it is cooled by helium gas instead of by water, it is free from the water-induced instabilities which caused the catastrophe at Chernobyl. Unfortunately, the HTGR never captured a substantial share of the nuclear power market. General Atomic sold one each of two versions of the HTGR. The first was a 40-megawatt (electric) version, which produced electricity for a utility company at Peach Bottom, Pennsylvania. It was turned off a long time ago because the utility decided it cost more to run it than 40 megawatts was worth.

Peach Bottom was always intended to be a small-scale experiment, a harbinger of bigger and better things to come. The second HTGR sold was eight times more powerful, a 300-megawatt version which is now running at Fort St. Vrain in Colorado. However, the Fort St. Vrain reactor had a technical problem. When you tried to run it at full power, the temperature in the core did not stay steady but wiggled a bit, probably because of some complicated coupling between the thermal expansion of graphite blocks in the core and the flow of the cooling gas through channels in the blocks. The temperature wiggles did not look dangerous, but to be on the safe side the reactor was licensed to run at only 70 percent of full power. At 70 percent power it ran smoothly. Still, you could not call it an outstanding success of HTGR technology. You could not expect other utility companies to come rushing with orders for more copies of the Fort St. Vrain reactor until this little problem was fixed. The problem was ultimately fixed, but too late. By the time it was fixed, the market for nuclear power reactors of any kind had collapsed.

So no more HTGR reactors were sold. In the meantime General Atomic had been bought and sold three times, but these upheavals left the company largely intact. General Atomic is still in business and some of the people there still have dreams of selling HTGR reactors. A few years ago the president of the company decided to hold a class reunion for the Class of 1956. He invited all the surviving members of the group of consultants who had started the company with high hopes in 1956 to come back and have another look at it. We had in the meantime grown old and dignified. We had become too important and too busy to come back for three months and work out new inventions. The most we could do was to come back for two days and have a good time remembering our lost youth. We did have a good time. And incidentally the General Atomic staff told us about their recent activities and about their plans for the future.

The main thing which the General Atomic people had to tell us was the result of two safety analyses of their full-scale HTGR power reactor. By full-scale they mean 1,000 megawatts electric, two and a half times the power output of Fort St. Vrain. For many years they had concentrated major effort on the detailed design of a full-scale 1,000-megawatt HTGR, a reactor which has not yet been built. In the meantime, two independent safety analyses of the full-scale HTGR have been done, one by a group of experts in the United States, the other by a group in Germany. Neither group of experts was connected with General Atomic; neither group had any commercial incentive to make the HTGR look good. And both groups came out with similar conclusions: in a certain well-defined sense, the HTGR is roughly a thousand times as safe as a light-water reactor of equal power.

The meaning of this statement is the following. The experts analyzed billion-year accidents, caused by combinations of stupidity and bad luck more extreme than anything we saw at Three Mile Island or at Chernobyl. A billion-year accident requires so much bad luck that it is supposed to happen only once in a billion years of reactor running time. A billion-year accident is a hell of a lot worse than Three Mile Island, and about a hundred times worse than Chernobyl. The reactor core vaporizes, the concrete containment building splits open, the atmosphere happens to have an inversion layer at the worst height, and the wind is blowing in the worst direction over a region of high population density. You do not need to believe in the accuracy of the calculation which says that this disaster happens once in a billion years. All that you need to believe is that it is possible to apply the rules of the accident-analysis game fairly, so that a billion-year accident for a light-water reactor and a billion-year accident for the HTGR are in some real sense equally unlikely. The results of the analyses are then startlingly favorable to the HTGR. The billion-year accident of a light-water reactor kills 3,300 people immediately and 45,000 people by delayed effects of radiation. The billion-year accident of the HTGR kills zero people immediately and seventy people by delayed effects. The numbers make no claim to accuracy, but the conclusion is qualitatively clear. It is conceivable that a mishandled HTGR may kill people, but it cannot kill them wholesale.

The next question that arises is then, if the HTGR is a thousand times as safe as a light-water reactor, and if public worries about accidents are threatening the very existence of the nuclear power industry, why is there not a crowd of utility executives standing at the door of the General Atomic sales office, waiting to trade in their light-water reactors for a shiny new HTGR? The answer to this question is simple. Even if the utility executives were in a mood to buy new and improved nuclear power plants, General Atomic would have none ready to sell. The full-scale HTGR has never been built. The components are not in production. The final stages of its engineering development are not complete. If a customer should now come to General Atomic wanting to order a full-scale HTGR, the best that General Atomic could do would be to say: "Well, wait a moment. If you can help us raise a half billion or so of government money to finish the engineering development, and if we don't run into any unexpected snags, with luck we could be ready to begin construction in a few years, and if the licensing goes smoothly you should have your reactor on line within ten years after that." This is not the kind of answer which brings utility executives running to place orders. Nobody in his right mind wants to plunge into a huge capital investment which will only begin to pay dividends, if all goes well, twelve years later. Only governments can afford to make such investments, and if they are wise they do not make them very often.

So here we see in a nutshell the tragedy of nuclear power in the United States. At the time when there was a market for power reactors, there was a company staffed by capable and dedicated people, with designs for a safer reactor, eager to go ahead with building it. And nothing could be done in less than twelve years. That is why I chose for the title of this chapter, "Quick Is Beautiful."

I told this story of the two reactors, the TRIGA which was finished and ready to go in three years and the HTGR which cannot be ready in less than twelve years, because I happen to have been personally involved with them. Similar stories could be told about many other industrial products. The nuclear industry is not the only one which has suffered from a hardening of the arteries and has lost the ability to react quickly to changing conditions and changing needs. The difference between a three-year and a twelve-year reaction time is of crucial importance. The rules of the game by which public life is governed, both in the United States and in the world outside, are liable to drastic and unpredictable change within less than ten years. By rules of the game I mean prices, interest rates, demographic shifts and technological innovations, as well as public moods and government regulations. We have recently seen some spectacular changes in the rules of the game which the automobile industry has to play. We can expect such sudden changes to occur from time to time, but nobody is wise enough to predict when or how.

Judging by the experience of the last fifty years, it seems that major changes come roughly once in a decade. In this situation it makes an enormous difference whether we are able to react to change in three years or in twelve. An industry which is able to react in three years will find the game stimulating and enjoyable, and the people who do the work will experience the pleasant sensation of being able to cope. An industry which takes twelve years to react will be perpetually too late, and the people running the industry will experience sensations of paralysis and demoralization. It seems that the critical time for reaction is about five years. If you can react within five years, with a bit of luck you are in good shape. If you take longer than five years, with a bit of bad luck you are in bad trouble.

Let me go back to the example of the nuclear reactor industry. What happened between 1956 and 1987 to cause such a disastrous slowing down of the reaction time? Part of the loss of flexibility can be blamed on government regulations and part can be blamed on the hardening of arteries in individual heads. The people who run the industry are not as young as they were. But regulation and aging are not the whole story. There are also some identifiable errors of policy which contributed to the slowing down. In my opinion, the two chief causes of the loss of flexibility of the industry were bandwagon-jumping and false economies of scale.

Bandwagon-jumping is not always bad. Only, before you jump on, you should look carefully to see whether the bandwagon is moving in the direction you want to go. If you are doubtful about the direction, it is a good idea to wait. In the case of the American nuclear power industry, the bandwagon was started by Admiral Rickover, who developed with admirable speed and efficiency a reactor to drive nuclear submarines. Rickover's reactor went into production and gave a flying start to the industry. It was a pressurized light-water reactor. So the light-water bandwagon started to roll. When the time came to build power reactors for the civilian utility market, everybody except General Atomic jumped onto Rickover's bandwagon. Unfortunately, they overlooked a well-known fact about submarines. There is not much room to spare in a submarine. Therefore the most important requirement for a submarine reactor is to be compact, to have a lot of power in a small volume. But when you build reactors for civilian utilities, the most important requirement should be safety. Other things being equal, the more compact a reactor is, the more power it generates in a given volume, the more quickly it will melt or vaporize in case of an accident. The shorter the time that is available before the reactor melts, the easier it is for somebody pressing the wrong switches to turn an accident into a catastrophe.

So compactness and safety are not running in the same direction. The main reason why the HTGR is safer than a light-water reactor is that it is less compact. What is good for submarines is not necessarily good for civilians. But once almost everybody had jumped onto Rickover's bandwagon, it became very difficult for anybody to jump off. By jumping on too soon, the nuclear industry deprived itself of alternative technologies which were leading in different directions. When the public became aware of the deficiencies of light-water reactors, the Rickover bandwagon ground to a halt, and the passengers were left stuck in the mud with nowhere else to go.

The effect of the bandwagon in immobilizing the nuclear industry was bad enough, but the effect was made much worse by a second factor, the pursuit of false economies of scale. I am not denying the reality of economies of scale. I am not recommending "Small is Beautiful" as a suitable motto for the petrochemical industry. Up to a point, big plants are usually more economical than small ones. Big nuclear reactors, up to a point, generate cheaper electricity than small ones. But the economy of scale is lost or even reversed when the big plant takes too long to build. If a plant takes ten years to build, it is almost certainly too big. The economy of scale is likely to be canceled out by interest charges and by loss of flexibility, and it will often happen that changes in the rules of the game make the big plant obsolete before it even comes on line. So you should pursue economies of scale up to the point where each unit takes about five years to bring on line, but no further. Further than that, it is a false economy. The light-water reactor industry probably made a fundamental mistake in going to 1,000-megawatt units. The expected economy of scale seems to have been illusory. Unfortunately, General Atomic felt compelled to make the same mistake with the HTGR. Just to keep up with the Joneses, General Atomic concentrated its efforts on the 1,000-megawatt monster which cannot be ready when it is needed.

The market for nuclear power reactors in the United States is at the moment nonexistent. Nobody knows whether the market will revive in the future. The hopes of the industry rest on the possibility that there will be some new oil crisis or some unpredictable change of political mood which will create a massive new demand for nuclear power. If this ever happens, the demand will be for reactors which are safe, and flexible, and quick to build. The 1,000-megawatt HTGR is safe but not quick. Perhaps General Atomic might finally achieve its rightful share of the market, if it could be ready when the time comes with a HTGR reactor of modest size, thoroughly tested and debugged, and capable of being mass-produced in a hurry.

One of the most beautiful pieces of technology I have ever seen is the factory in Everett north of Seattle where they build Boeing 747s. The Boeing 747 is not small and neither is the factory. But the factory is wondrous quick. At the time I visited, they were turning out 747s at the rate of one a week. That is the sort of operation which I have in mind when I say, "Quick is Beautiful."

People of my generation who lived through World War II have vivid memories of monumental confusion and incompetence—after all, the word "snafu" is of World War II vintage—and in spite of all that, we remember that in the end things got done. When Winston Churchill became prime minister in 1940, England was desperately short of ships, airplanes, tanks, guns, everything that we needed to fight a war. I saw how bad things were when the little old 22-caliber rifles that the boys in my school at Winchester used for target practice were taken away from us and given to the army. Those rifles probably last saw active service in the Crimea in 1856. In 1940 Winston Churchill spoke on the radio and said, "I am sorry I cannot do anything for you in less than three years. I give an order to build a factory today, and in two years you have nothing, in three years you have a little, in four years you have a lot, in five years you have all you want." He was right. In five years we had all we wanted and in five years the war was over.

These experiences of World War II made an indelible impression on people of my generation. At the bottom of our hearts we still believe you can have anything you want in five years if you need it badly enough and if you are prepared to slog your way through the barriers of confusion and incompetence to get it. Some of us even believe that if tomorrow the millions of unemployed would do us the favor of starting a bloody revolution, the shock to our system would be sufficient to push us into a serious public-works program, and we would end up within five years with a functioning full-employment economy. Such ideas are totally contrary to the accepted wisdom of our economists and politicians. The accepted wisdom says that, no matter what we decide to do about economic problems, we cannot expect to see any substantial results before the year 2000. The accepted wisdom is no doubt correct, if we continue to play the game by the rules of today. But anybody who lived through World War II knows that the rules can be changed very fast when the necessity arises.

Why is it that our whole economic and political system has tended recently to become so sluggish and inflexible? Why have we become resigned to the idea that nothing substantial can ever be done in less than ten years? Obviously there are many reasons. But I believe the principal reason for this sluggishness is that our whole society has fallen into the same trap as our nuclear industry. Not only in the nuclear industry but in many other industries and public institutions, we have pursued economies of scale which turned out to be false. One of the most fundamental false economies of scale is the overgrowth of cities. At one time it looked economically attractive to cram millions of people together into huge agglomerations. The biggest cities everywhere are running into social problems which indicate that this was a false economy.

When we turn from sociology to biology, we see the same historical processes at work. So long as no sudden changes in the rules of the game occurred, all through the soft swampy sluggish 100-million-year summer of the Mesozoic era, the dinosaurs pursued their economies of scale, growing big and fat and prosperous, specializing their bodily structures more and more precisely to their chosen ecological niches. Then one day, as we learned from the observations of Luis Alvarez and his colleagues at Berkeley, a comet fell from the sky and covered the Earth with its debris. The rules of the ecological game were changed overnight, and our ancestors, the small, the quick, the unspecialized, inherited the Earth.

Let me now tell you a more cheerful story. In Princeton we have had two projects, each of them in its own way trying to contribute to a solution of the energy problem. The two efforts stood side by side on the Forrestal Campus of Princeton University. One of them is the TFTR, the Tokamak Fusion Test Reactor, the white hope of the magnetic confinement fusion program, a magnificent piece of engineering, lavishly funded by the Department of Energy. It cost $300 million and has now been in operation for several years. It is exploring the technology for commercial fusion reactors which will possibly begin producing electricity fifteen or twenty years later.

The other project, the one with which I have had the honor to be associated, was the Princeton Ice Pond. The 1980 version of the ice pond was a square hole in the ground with a dirt berm around it and a sheet of Griffolyn plastic lining its bottom. Two men with a mechanical digger dug the hole in January 1980. We rented a commercial snow machine and squirted snow over the hole during the cold days and nights of February, until we had something that looked like the Disneyland Matterhorn. Halfway through the snowmaking, we found out that we didn't need that fancy ski-resort snow machine. We didn't need skiing-quality snow for our pond. We found out that for our purposes a fireman's fog-nozzle which we could buy for $300 would do the job well enough. Our Matterhorn stood high and proud for a few weeks. Then the March sun shrank it down a bit, and the April rains reduced it to a pool of slush, filled up to the top of the berm.

We covered it over with an insulating layer of plastic and straw, and on top of that we put an air-supported mylar dome to keep the straw dry. In June a hefty hailstorm wrecked the mylar and so we made do with wet straw for the insulation. I am not claiming credit for any of this. The project was run by Rob Socolow, Don Kirkpatrick, Ted Taylor, and their students at the Center for Environmental Studies of Princeton University. I was only an unskilled laborer who went out to help them occasionally. In June we measured the contents of the pond and found that we had about 450 tons of ice with some water underneath it.

All through an exceptionally hot Princeton summer we successfully air-conditioned a building by circulating fresh water from the bottom of the pond. We were melting ice at a peak rate of about 7 tons a day—beautiful cool white ice with crevasses into which we could descend and enjoy Alpine scenery under the blazing Princeton sun. When the hot weather came to an end at the beginning of October there were still about 150 tons of ice left.

The 1980 ice pond proved that the idea works, but it was a muddy and messy job, suitable for Princeton students but unsuitable for suburban homeowners or business executives to look at out of their picture windows. The objective of our later experiments has been to make the ice pond elegant. We want also to make it out of more permanent materials, so that it will not need too much maintenance. The ideal is to have a system that works year after year, so that the owners can forget about it. Above all, the pond must not give the owners headaches. People have enough headaches already. If ice ponds cause additional headaches, nobody will want to own them. We need to demonstrate that the job can be done without creating an eyesore, in a style compatible with the aesthetic standards of real estate developers and architects.

The ice-pond project was not supported by the Department of Energy. We applied to DOE for funds several times, but the most DOE could do for us was to tell us to apply to

Housing and Urban Development, and when we applied to HUD they told us, not unexpectedly, to go back to DOE. The project finally got started in January 1980 because the Prudential Insurance Company decided we might be a good investment. The Prudential was prepared to spend $300,000 (not million) to find out whether we were as crazy as we looked. The Prudential did not require us to spend three quarters of the money on paperwork.

Why should the Prudential Insurance Company be interested in supporting a technologically primitive project like ours? It happened that the Prudential was investing its surplus cash in the construction of several office buildings in an industrial park near the campus. The possible payoff for the Prudential was a solar heating and cooling system for their office buildings. If our wildest dreams had come true, we would have been able to supply solar heating and cooling to a building at a capital cost equal to the cost of fuel and electricity used by an equivalent conventional system in about two years. In other words, the Prudential would have written off the cost of the solar system in two years and would enjoy free heating and cooling thereafter as long as the system lasted. The Prudential was prepared to gamble $300,000 on the remote chance that something like that might happen. Nothing like that has yet happened. Ice ponds are still experimental, unreliable and inconvenient. A lot of work has to be done before they will be a standard commercial product.

The key to cheap and reliable solar energy is to have a cheap and massive storage of heat and cold, massive enough so that it can ride over the annual weather cycle, heat being collected in summer and used in winter, cold being collected in winter and used in summer. The system which we had in mind for the Prudential building would use two ponds for storage, a hot pond containing 100,000 tons of hot water (roughly 2 acres, 30 feet deep), and a cold pond containing 10,000 tons of ice (roughly a quarter of an acre, 30 feet deep). We started first with the ice-pond experiment because the money came through in January 1980 just in time for the snowmaking. It is easier to make snow in a hurry in winter than to make hot water in a hurry in summer.

The beauty of solar-pond technology lies in the fact that your mistakes do not stay hidden for long. We made a mistake with the mylar dome; in two months the hail ripped it apart and we were ready to try something else. We made another mistake with our heat engine. An important part of our original plan was to generate homemade electricity with a heat engine, using the hot and cold ponds as heat source and heat sink. The ideal efficiency of an engine working between 140 degrees Fahrenheit and ice temperature is 20 percent, so we thought we could expect to run a real engine making electricity at 10 percent efficiency. We found a supplier in Florida who claimed he could sell us an engine for $12,000, putting out 10 kilowatts of electricity; $1,200 capital cost per kilowatt, and zero cost for fuel, would compete well with central-station power.

The Department of Energy people told us, in their lordly fashion, that the Florida engine was a pile of junk, but that only made us more determined to prove the Department of Energy wrong. We arranged with the designer in Florida to rent one of his engines for three months. He drove it himself all the way from Florida to Princeton and proudly handed it over. The machine looked good, not like a pile of junk at all. The designer told us how he had run it in Florida using muddy swampwater full of frogs for the cold end, and the machine handled the frogs smoothly without ever getting choked up. The whole deal looked good. We paid our three months' rent with the option to purchase, and handed the machine over to Greg Linteris, one of our undergraduates, to measure its performance carefully. A few weeks later Greg Linteris reported his results. Unfortunately, it turned out that the designer of the machine did not understand three-phase circuitry. All his numbers for electric power were too high by a factor of the square root of three. He was better at handling frogs than at handling complex numbers. The actual efficiency of the machine was 6 percent and not 10. So we sadly shipped it back to Florida.

The Prudential Insurance Company has never been enthusiastic about hot-water ponds and heat engines. They are happy to buy their electricity from Public Service Electric and Gas, and have no wish to get into the utility business themselves. So it was easy to agree with them to go ahead with the ice-pond experiment and drop the heat engine for the time being. If all goes well with the ice ponds, we can come back to heat engines later.

After the first experiments in Princeton were finished, my friend Ted Taylor started a commercial company to develop and build ice ponds. The company has not made him rich, but it has taught him a lot about the difficulties of selling a new technology. Up till now he has succeeded in selling and building only two functional ice ponds. His first commercial customer was the Kutter Cheese Company, a small food-processing business in the farm country of western New York State. Food-processing plants are an easier market for ice ponds than residential or office air conditioning. A cheese factory has a predictable demand for refrigeration all the year round, and it does not need to look beautiful. The Kutter Cheese Company is happy with its ice pond and is saving some of the money which it formerly spent on electricity for refrigeration.

The second of Taylor's commercial projects is an ice pond at Greenport on Long Island. The Greenport ice pond is quite different from the others. It is made by freezing salt water taken directly from the Atlantic Ocean. After the big pile of salt-water snow has been left to settle for a few weeks, almost all the salt has trickled down to the bottom and the bulk of the remaining ice is remarkably pure. The ice in the middle of the pile is so pure that it easily complies with the New York State standards for public supplies of drinking water. It happens that Long Island suffers from a shortage of available fresh water, and that is why the village of Greenport became interested in ice ponds. We are hoping that before long the water from the

Greenport ice pond may be replenishing the reservoirs of eastern Long Island. It turns out that the process of rapid freezing and slow thawing is effective not only for removing salt from seawater but also for removing most of the disagreeable man-made chemicals. The ice-pond technology may in the end find its major application in water purification rather than in refrigeration. As usual when a new technology is being developed, the most serious obstacles are institutional rather than technical. How do you fit ice ponds into the regulations governing the specifications of municipal water supplies? Within a few years we will find out whether the need for a new technology is acute enough to overcome the institutional obstacles.

That is the story of the Princeton Ice Pond. I told the story because it illustrates what I have in mind when I ask for a technology with a quick response. I do not claim that ice ponds will solve any big social problems, or that the little game we have been playing in Princeton has demonstrated the existence of an economically viable ice-pond technology. I claim only that ice ponds are an example of a technology free from the rigidities and the decade-long delays which have made both fission and fusion power unable to respond to public need. Ice ponds may or may not turn out to be cheap and effective. If they fail, they will fail quickly and we shall not have spent half a lifetime proving them useless. If they succeed, there is a chance that ice ponds and solar hot-water ponds could be deployed rapidly on a large scale. Sites could be surveyed, holes in the ground dug, and plumbing fixtures installed by thousands of local contractors responding to local demand. Plastic liners and pipes and solar collectors could be mass-produced in factories. A whole new industry could grow up in a few years like the industries of World War II.

All this is only a dream, or at best a remote possibility. But there is no reason why a new technology has to develop like fission and fusion on a thirty-year time scale. All it needs in order to go fast is small size of units, simple design, mass production, and a big market. When I went to the Forrestal

Campus outside Princeton and saw those two machines, the $300 million TFTR and our little ice pond, what I saw in my mind's eye was a dinosaur and an early primate. I wonder how long it will be before the next comet falls.

Finally, I come to the subject of genetic engineering. When I compare the biological world with the world of mechanical industry, I am impressed by the enormous superiority of biological processes in speed, economy and flexibility. A skunk dies in a forest; within a few days an army of ants and beetles and bacteria goes to work, and after a few weeks barely a bone remains. An automobile dies and is taken to a junk yard; after ten years it is still there. Consider anything that our industrial machines can do, whether it is mining, chemical refining, material processing, building or scavenging; biological processes in the natural world do the same thing more efficiently, more quietly and usually more quickly. That is the fundamental reason why genetic engineering must in the long run be beneficial and also profitable. It offers us the chance to imitate nature's speed and flexibility in our industrial operations.

It is difficult to speak of specific examples of things genetic engineering may do for us. Specific examples always sound like stories out of *Astounding Science Fiction* magazine. Here are three long-range possibilities. First, the energy tree, programmed to convert the products of photosynthesis into conveniently harvested liquid fuels instead of cellulose. Second, the mining worm, a creature like an earthworm, programmed to dig into any kind of clay or metalliferous ore and bring to the surface the desired chemical constituent in purified form. Third, the scavenger turtle with diamond-tipped teeth, a creature programmed to deal in a similar fashion with human refuse and derelict automobiles. None of these creatures performs a task essentially more difficult than the task of the honeybee with which we are all familiar. But it is a sound instinct which leads us to be distrustful of such grandiose ideas. If we pursue long-range objectives of this kind, we are likely to find ourselves involved in a twenty-year development program with all the inertia and inflexibility of a nuclear power program. The whole advantage of biological technology will be lost if we let it become rigid and slow.

I hope we shall make our entry into genetic engineering in a pragmatic and opportunistic fashion, choosing projects which lead quickly to short-range objectives, choosing processes which fit conveniently into the framework of existing chemical and pharmaceutical industries. That is the way we went into nuclear engineering with the TRIGA reactor in 1956. And that is the way we should have continued in nuclear engineering if we had been wiser. Above all, we should try to exploit the small scale and fine-tuning of biological processes to achieve production facilities which are rapidly responsive to changing needs. Never sacrifice economies of speed to achieve economies of scale. And never let ourselves get stuck with facilities which take ten years to turn on or off. If we follow these simple rules, there is a good chance we will help genetic engineering to fulfill the promise of a cleaner and more livable world for mankind, the promise which nuclear energy once made but was never able to fulfill.

In my discussion of nuclear energy, I spoke only about reactors and not about bombs. In my opinion, the biggest mistake of the nuclear scientists was their enthusiastic pursuit of bombs. I cannot consider Three Mile Island to be an event in any way comparable with Hiroshima. The question then has to be faced, whether the pursuit of genetic engineering might expose us to dangers comparable with the dangers of nuclear weapons and nuclear war. I believe the answer is no, with one essential proviso. The proviso is that the existing laws restricting experimentation on human subjects continue to be enforced. In other words, genetic engineering must stop short of monkeying around irresponsibly with the species Homo sapiens. So long as Homo sapiens is left out of it, I do not see how genetic engineering can lead to military abuses significantly worse than the old-fashioned chemical and biological weapons with which we are unhappily familiar. But do not take my word for it. Expect the unexpected. Keep a careful lookout for dangers ahead, and when something on the horizon looks bad, call a halt, blow the whistle, and try to find a different way to go. If the worst comes to the worst, there are other ways to make a living.

I do not think that the theoretically possible dangers of genetic engineering will turn out to be real. I think that the benefits of it will be large and important. I wish luck and success to the young scientists who are now beginning their careers as genetic engineers. The best luck that I can wish them is to have as much joy with genetic engineering as we had with the TRIGA reactor and with the Princeton Ice Pond. Developing a new technology is hard work, but it is also fun. If they are lucky, they will find, as Michael Pupin found a hundred years ago, that invention is just as creative and just as exciting a way of life as scientific discovery. When they grow old they may find, as Michael Pupin found, that the life of an inventor also provides ample room for philosophical reflection and for active concern with the great problems of human destiny.

9

SCIENCE AND SPACE

Space science happens to be my own professional turf. I am taking for granted that you share my enthusiasm for the beauty and richness of this little piece of turf, which is only a small piece of the kingdom of science. I will be talking mostly about American experiences and American problems. Please forgive me if I do not all the time keep reminding you that science is international, that science is only a small part of human culture and that American space science is only a small part of science. I am giving you a parochial view of a great human enterprise.

Science has never been the main driving force of the American space program, and the space program has never been the main driving force of science. That is as it should be. Science and space each have their own objectives and grand designs, independent of each other. Science is at its most creative when it can see a world in a grain of sand and a heaven in a wild flower. Heavy hardware and big machines are also a part of science, but not the most important part. The American space program is at its most creative when it is a human adventure, brave men daring to ride their Moon Buggy over the foothills of the Lunar Apennines to the brink of the Hadley Rille. Precise observations and dating of moon rocks are also a part of space exploring, but not the most important part.

The main driving forces of the space program have been political, military, and commercial rather than scientific. If we measure the size of programs by total effort and budgetary outlay, then roughly 10 percent of the space program is science and roughly 10 percent of the science program is space. Nevertheless, the 10 percent area of overlap between science and space is of vital importance to both parties, and since I am a scientist I shall concentrate on this area of overlap. I shall survey the highlights of space science during the last thirty years and try to derive some useful lessons for the future.

There are two ways of looking at space science. One may approach it either from the space side or from the science side. If one comes from the space side, it is natural to adopt a mission-oriented approach, measuring success and failure by missions done and not done. If one looks at the last thirty years from the mission-oriented point of view, one sees a large number of splendid successes, a smaller number of sadly missed opportunities, and a very few outright failures. This is the point of view of the space professionals, and it is also the point of view of the general public in so far as the general public is interested in space science at all. I am not saying that the mission-oriented approach is wrong. But it is not the whole story. Since I come from the science side, it is natural for me to look at space science in a different way, with a science-oriented approach.

The science-oriented approach measures success and failure of missions by looking at the quality of their scientific output. It sees space science embedded in a wider context of ground-based science. It asks of each mission not merely the easy question, "Did it work?" but the more difficult questions, "So what?" "What did we really learn?" "Was that the right thing to observe?" "Was that the quickest, or the cheapest, or the most effective way to make the observation?" The science-oriented approach does not believe in pass-fail grading. Scientifically speaking, there is such a thing as total failure of a mission, but there is no such thing as total success. A successful mission will raise new questions as often as it answers old ones.

I like to deal in particular instances rather than in generalities. Let me begin with a concrete example, a successful space mission which I know something about, since it was conceived and operated in Princeton. This is the orbiting ultraviolet telescope called Copernicus. It was launched in 1972, just in time to celebrate Copernicus's five hundredth birthday. From a mission-oriented point of view, Copernicus was a brilliant success. It did exactly what it was designed to do, taking high-resolution ultraviolet spectra of hot stars and measuring absorption lines produced by atoms and ions in the interstellar gas. It was supposed to last for a year and actually lasted eight. It kept on producing data, year after year, until it finally died of old age. People at Princeton are still working on the data and still publishing papers about it. The Princeton astronomers are justly proud of their Copernicus. They invented it, designed it, fought for it, used it, and nursed it through its declining years. A remarkable achievement for a small university department with a little help from NASA.

But when one looks at Copernicus from the science-oriented point of view, the picture is more complicated. The original idea of Copernicus arose in the 1950s in the mind of Lyman Spitzer. Spitzer was, and still is, a pioneer in exploring the nature and distribution of the interstellar gas in our galaxy. In the 1950s the main evidence for the chemistry of the interstellar gas came from narrow absorption lines of sodium and calcium seen in the optical spectra of certain stars. Why sodium and calcium? Because sodium and calcium are the only elements which have absorption lines in the right part of the optical spectrum to be seen with ground-based optical telescopes. But sodium and calcium are minor constituents of the gas, so that the sodium and calcium lines do not give good information about the behavior of the gas in general. The majority of atoms in the gas belong to the common elements hydrogen, carbon, nitrogen and oxygen, which have absorption lines only in the far ultraviolet. So Spitzer decided it would be a good idea to put into orbit a far-ultraviolet telescope able to record and measure accurately the absorption lines of the abundant elements in the interstellar gas. NASA agreed that this was a good idea, and Spitzer's telescope was approved in 1960 as the third in the series of Orbiting Astronomical Observatories. The contract for its construction was signed in 1962, with launch scheduled for 1965.

For various reasons, partly technical and partly political, the launch of Copernicus was delayed by seven years. This meant that a telescope designed to answer the scientific questions of the 1950s was launched in the 1970s. Between the designing and the launching of Copernicus, a revolution occurred in radio astronomy. Radio astronomers observing from the ground learned how to see the interstellar gas with millimeter waves. Millimeter-wave astronomy answered the main questions about the chemical composition of the gas. Millimeter-wave telescopes on the ground did a large part of the job Copernicus was designed to do, more quickly and more cheaply and more comprehensively. This does not mean that Copernicus was scientifically useless. The Copernicus observations complement the millimeter observations nicely, giving information especially about the dilute high-temperature component of interstellar gas which is invisible to radio telescopes.

Copernicus is far from being a scientific failure. But it was not the instrument astronomers would have chosen for answering the exciting scientific questions of the 1970s. In the seventies we needed an ultraviolet telescope with which we could study the newly discovered X-ray sources, quasars, and other mysterious objects in which violent dynamical processes are occurring. All the newly discovered objects are faint. And Copernicus, having irrevocably chosen to sacrifice light-gathering power for the sake of high spectral resolution, could not see faint objects. The only time Copernicus got a chance to look at an exciting new object was when Nova Cygni flashed in the Northern sky for a few nights in August and September of 1975.

Copernicus was only one of many successful missions in the scientific part of the American space program. I spoke about Copernicus in detail because I believe it illustrates a general problem which recurs over and over again in the history of space science. The problem is the mismatch in time scale between science and space missions. The cutting edge of science moves rapidly. New discoveries and new ideas often turn whole fields of science upside down within a few years. The discovery of pulsars in 1967 burst on the astronomical scene as suddenly and unexpectedly as Nova Cygni, and transformed within a year the way we think about the late phases of stellar evolution. The effect of such discoveries is to change the priorities of science, to change the questions we want to have answered. Every young scientist's dream is to be able to say what the nineteen-year-old mathematical genius Évariste Galois said in 1830, "I have carried out researches which will halt many savants in theirs." Science must always be ready to halt and switch its objectives at short notice. To make this possible, the tools of science should be versatile and flexible.

Unfortunately, flexibility and versatility are hard to achieve in space missions. In the space program, plans for missions and designs of instruments tend to be frozen many years in advance. Copernicus, with its twelve-year interval between design and launch, was perhaps an extreme case. But intervals of eight and ten years are not uncommon. In most major space missions, the instruments were designed to answer the questions which seemed important to scientists a decade earlier. The bigger and more ambitious the missions become, the more difficult it is to reconcile the time scale of the missions with the time scale of science. Space science begins to look like a two-horse shay with a carthorse and a racehorse harnessed together.

What should the space program do to recapture its lost youth? Later on I will make some specific suggestions of things we might do to rejuvenate space science. But first I will go back again to the past. It is useful to look at the past, so as to learn from the mistakes of the past how to do better in the future, and also to learn from the successes of the past how not to do worse in the future.

The thirty years since Sputnik divide themselves conveniently into two periods, Apollo and post-Apollo. The Apollo period ends with the departure of Harrison Schmitt and Eugene Cernan from the Moon in December 1972. The Apollo period is particularly instructive because we can see clearly with the benefit of hindsight which parts of the space enterprise in those years were scientifically the most productive. In the Apollo period there was a strong negative correlation between budgetary input and scientific output. The negative correlation was neither planned nor expected. It just happened because science is unpredictable. The most expensive missions produced the least significant science, and the cheapest missions produced the most exciting science.

The space program of the Apollo period included three main types of exploratory mission: First, there were manned missions culminating with Apollo; second, unmanned planetary missions culminating with the Mars Mariners; and third, the series of X-ray sounding-rocket missions culminating with the launch of the first X-ray satellite Uhuru. The costs of the Apollo missions, the Mariner missions and the X-ray missions were roughly in the ratio of a hundred to ten to one. We cannot attach numerical values to the scientific results of the various missions. The relative value of different types of scientific information must be to some extent a matter of personal taste. Nevertheless, I am prepared to say unequivocally that the beginning of X-ray astronomy, opening up a new window into the universe and revealing the existence of several new classes of astronomical object, was the most important single scientific fruit of the whole space program.

The newly discovered X-ray sources gave an entirely fresh picture of the universe, dominated by violent events, explosions, shocks and rapidly varying dynamical processes. X-ray observations finally demolished the ancient Aristotelian view of the celestial universe as a serene region populated by perfect objects moving in eternal peace and quietness. The old quiescent universe of Aristotle, which had survived essentially intact the intellectual revolutions associated with the names of

Copernicus, Newton and Einstein, disappeared forever as soon as the X-ray telescopes went to work. And the new universe of collapsed objects and cataclysmic violence originated in the cheap little sounding rockets of the sixties, popping up out of the Earth's atmosphere and observing the X-ray sky for only a few minutes before they fell back down. The most brilliant achievement of the sounding-rocket era was Herbert Friedman's 1964 measurement of the angular size of the X-ray source in the Crab Nebula using the Moon as an occulting disk.

An occulting disk usually means a round piece of metal which an astronomer puts onto a telescope to cut out the light of the sun when the telescope is used to observe the sun's corona. Herbert Friedman wanted to find out whether the newly discovered X-ray source in the Crab Nebula was extended over the volume of the nebula or concentrated in a single star. He knew that the moon would pass exactly in front of the nebula on July 7, 1964. He arranged to launch his X-ray sounding rocket just before the moon occulted the nebula. If the X-ray source was point-like, the rocket would see the X-rays cut off abruptly. If the source was extended, the cutoff of X-rays would be gradual. The result of the observation showed that the source is extended. The X-rays come mainly from energetic electrons circulating within the nebula.

The Aerobee rocket which carried Friedman's X-ray detectors gave him only 5 minutes of observing time from above the atmosphere. The occultation of the Crab Nebula by the Moon occurs only once every 9 years. Friedman and his rocket were ready when the time came, so that the flight of the Aerobee coincided with the central 5 minutes of the occultation. Friedman was able not only to prove that the X-ray source was extended but also to measure its size. He found that the source has a diameter of 1 light-year, significantly smaller than the 3 light-year diameter of the visible nebula. Friedman's observation was a classic example of economy in the use of scientific resources. The cost of X-ray astronomy in the Apollo period was less than 1 percent of the total budget for space.

The manned missions which absorbed the bulk of the space budget in those days yielded a harvest of solid scientific information about the Moon. Samples of various types of moon rock were brought home and analyzed and dated. The stratigraphy of the Moon was clarified and its early history elucidated. Its seismic and magnetic characteristics were measured. All this was good science. But it was not great science. For science to be great it must involve surprises, it must bring discoveries of things nobody had expected or imagined. There were no surprises on the Moon comparable with the X-ray burst sources or with the X-ray binary sources which gave us the first evidence of the actual existence of black holes in our galaxy. Everything discovered on the Moon could be explained in terms of conventional physics and chemistry.

However, God played us a joke which made the Apollo program scientifically worthwhile. Because NASA was interested in bringing back rocks from the Moon, NASA was also interested in funding meteoritic science in a rather lavish fashion. We had to have good instruments in order to analyze the moon rocks. So a whole generation of meteoritic chemists was supplied with the necessary instruments, good mass spectrometers and microchemical apparatus of various sorts, and there they were with this beautiful apparatus, waiting for the rocks to come back from the Moon. And at just that happy moment, God decided to play a hand and threw down in Mexico a piece of rock which was more interesting than all the moon rocks combined, namely, the Allende meteorite. It was not only more interesting, it was also bigger. It weighed more than twice as much as all the moon rocks put together, and that meteorite contained a wealth of wonderful things, isotopic anomalies which gave evidence of pre-solar composition of the meteorites, evidence for things that you can actually hold in your hand which are older than the Sun, evidence of events which took place before the solar system was formed. All these things came from the chemical analysis of micro-inclusions in the Allende meteorite. All that we got more or less for free.

The unmanned planetary missions of the Apollo period were intermediate both in cost and in scientific importance between the manned missions and the sounding rockets. They were less costly than Apollo and less exciting scientifically than X-ray astronomy. The most exciting aspect of the planetary missions was their technical brilliance. They gave us celestial billiards, a new game which Giuseppe Colombo invented. The Mariner 10 spacecraft bounced around in the inner solar system, encountering Venus and Mercury repeatedly, taking pictures of Mercury from various angles. It was a spectacular demonstration of skill. The Mariner missions gave us some beautiful surprises, high temperatures and pressures and absence of water in the atmosphere of Venus, giant volcanoes and canyons, ancient craters and absence of canals on Mars. But the surprises were not of such a magnitude as to cause a scientific revolution. The newly discovered features of Mars and Venus were mysterious but not wholly unintelligible. The Mariner observations were a big step forward in the understanding of the planets. They were not the birth of a new science.

So much for the Apollo period. If space exploration had stopped at the end of 1972, we might have concluded that a simple mathematical law governs the scientific effectiveness of missions in space, namely, that the scientific output varies inversely with the financial input. If this law held good universally, the administration of space-science programs would be a simple matter. Just cut the budget and watch the science improve. But unfortunately, this simple managerial method does not always work as it should. The history of space science in the post-Apollo period showed a more complicated pattern.

In the seventies we again had three programs continuing the work begun in the sixties, the Skylab and Shuttle missions taking over from Apollo, the Viking and Voyager missions taking over from the Mariners, and the Einstein X-ray Observatory taking over from Uhuru. It was still true that the X-ray observations were first in scientific importance. The Einstein Observatory during its sadly short operational life poured out a steady stream of revolutionary discoveries, including the discovery of X-ray variability of quasars on a time scale of hours. The rapid variation of quasars implies that we have in some of these objects a switch which can turn the energy output of 10 billion suns on and off within an hour or two. Some switch! The X-ray telescope allowed us for the first time to penetrate close to the central core of the mysterious engines which drive these most violent objects in the universe. It was still true in the seventies that the X-ray discoveries were of greater fundamental importance than the planetary discoveries, even though the Viking and Voyager missions gave us a wealth of new scientific surprises as well as pictures of incomparable beauty. It was still true in the seventies that the planetary missions outstripped Skylab in scientific value.

But in the seventies, unlike the sixties, there was no longer a factor of ten difference in costs between the three types of mission. All three types of mission had become comparably expensive. Einstein was a little cheaper, but not enormously cheaper, than Voyager; Voyager was a little cheaper, but not enormously cheaper, than Skylab. By the end of the seventies, we could no longer say as confidently as we could in the sixties that the smallest and cheapest parts of the space program were scientifically the best. All parts of the program, irrespective of their scientific merit, had come to be dominated by large and expensive missions. And the program thereby lost the flexibility that the small missions of the Apollo period had kept alive.

Large missions have two outstanding defects which are apt to lead to scientific trouble. The first defect I already mentioned, the long lead-times which make large missions inflexible and unable to respond to new ideas. The second defect is the tendency of big missions to become one-of-a-kind. This defect is related to the political climate within which large missions must be presented to the government and to the public. In order to secure funding for a large scientific mission, the proponents are forced to talk about the important scientific problems which that mission by itself will solve. Then, in order to stay honest, they are forced to conform the design of the mission to their promises. The mission then becomes a one-shot affair, designed and announced to the public as the final answer to some big scientific question. The most unhappy example of this "one-shot" syndrome was the Viking mission, which was forced by the political circumstances of its origin to accept the impossible scientific task of deciding all by itself whether or not life exists on Mars.

If one looks in detail at the Viking experiments, it is difficult to imagine any combination of results which would have definitely proved or definitely disproved the existence of life on Mars, unless we had been lucky enough to find a cactus bush or an armadillo sitting immediately in front of the television cameras. One-shot missions are not a good way to do science. If we want to investigate seriously the question of life on Mars, the best way would be to plan a regular series of Mars missions, each one far less ambitious and elaborate than Viking, so that we could learn from the results of one mission the right questions for the next mission to ask. We could also learn from the mistakes of one mission how to avoid mistakes on the next. In almost any field of space science, whether we are exploring planets or galaxies or our own Earth, a series of modest missions is more likely than a single big spectacular to produce important discoveries.

The baleful effects of the one-shot syndrome can be seen not only in the Viking mission but also in X-ray and optical astronomy. The Einstein X-ray Observatory was magnificently productive while it lasted, but it was a one-shot performance with no follow-on. We must now wait many years before another mission can be launched to answer the new questions which Einstein raised. We would have been far better off scientifically with two or three small Einsteins in sequence instead of one big one. I must also confess that I am uneasy about the scientific justification of the Hubble Space Telescope, the big optical telescope which is to be launched into orbit in 1988. It is heresy or treason for a scientist to express misgivings about the Hubble Telescope. But I have to say in all honesty, the Hubble Telescope is a basket with too many eggs riding in it. It would have been much better for astronomy if we had had several 1-meter space telescopes to try out the instrumentation and see how the sky looks at a tenth-of-a-second-of-arc resolution, instead of being stuck with a single one-shot 2.4-meter telescope for the rest of the century. Perhaps I am being unduly pessimistic. The Hubble Telescope can hardly fail to make big discoveries when it comes into service in the late 1980s. But I have a sneaking fear that it may end up in the 1990s as Copernicus did in the 1970s, technically a glorious success but scientifically twenty years behind the times.

What have we done recently with smaller and less cumbersome space-science missions? In the last ten years we have had two magnificent missions, the IUE, International Ultraviolet Explorer, and the IRAS, Infra-Red Astronomy Satellite, both of them international projects with the United Kingdom and the Netherlands making major contributions. The IUE is an orbiting ultraviolet telescope which can look at a far greater variety of objects than Copernicus. Instead of looking at a few bright stars with very high spectral resolution, it looks at a large number of fainter objects with lower resolution. The IRAS is an orbiting infra-red telescope which produced the first complete survey of infra-red sources in the sky. Those two missions have been outstanding in their scientific output and intermediate in their level of cost. I wish we had more such missions. We desperately need more such missions.

If you go and look at the Goddard Space Flight Center in Maryland where the astronomer actually sits at the console and points the IUE telescope and takes the observations, it is wonderfully direct. She sits there and points the telescope and takes the data herself. It is all done in real time. She can decide what to do next, and if the spectrum is not particularly good, she can stop the observation and turn to something else. It is a beautiful system, small enough and informal enough so that people can get access to the telescope easily. It works. And it is the most productive telescope in existence, if you measure productivity as the NASA bureaucrats measure it, by pages per year in the *Astrophysical Journal* and other professional journals. That is a terrible way to measure scientific output, but I do not know how else to measure it. If you measure output by pages published per year, then this little telescope in the sky, the IUE, is twice as productive as any other telescope in the world. The IRAS lived and worked for only ten months before its supply of liquid helium ran out, but it was also an amazingly productive mission. That was the last of our successful intermediate-cost missions. Now let me turn to the future.

The Hubble Space Telescope is only one item in the plans for space science in the coming decade. The plans are subject to great political uncertainties. A large number of ambitious missions have been proposed and recommended by various committees of distinguished scientists. Few have been officially approved and funded. We have been suffering from a surfeit of committees. Committees do harm merely by existing. They run into the same trap, whether they are making recommendations about astronomy or high-energy physics or computers or nuclear power or plasma physics. What happens when you have a committee? Inevitably it concentrates its attention on the big items. The agenda is dominated by the biggest and most costly projects. No matter what the intention was, the big items receive the most attention and are emphasized in the final report. The Field Committee was a committee of astronomers which worked hard to draw up a ten-year plan for American astronomy. The Field Committee said that we need small missions, but the big missions have far more space devoted to them in the committee report. The big missions are overemphasized. When the experts testify to congressional committees and ask for funds, all the conversation is about the big items. In the old times, when missions were more numerous, when we had smaller instruments and more frequent flights, the program was less often in political trouble. We should take a lesson from the Japanese, who are now flying one space science mission every year. Their resources are limited, their spacecraft are small, and they are doing good science.

Three American and European missions have been approved and scheduled to fly, namely, Hubble Telescope, Galileo and Hipparcos. Hubble Telescope and Galileo are Shuttle missions, both stretching the limits of budgetary and political feasibility. Both have been subject to long delays and technical uncertainties resulting from difficulties in the development of the Shuttle. Hubble Telescope is a general-purpose optical instrument designed to give images about twenty times sharper than the best images obtainable from ground-based telescopes. It will explore the fine details of selected objects, mostly very dim and distant objects which cannot be effectively observed from the ground. Galileo is a planetary mission which will do for Jupiter what Viking did for Mars, sending probes deep into Jupiter's atmosphere and providing fairly complete photographic coverage of Jupiter and its satellites. Galileo, like the Hubble Telescope, is a one-of-a-kind mission. No further large missions to Jupiter are planned before the end of the century. If, as is likely, Galileo raises important new questions, we will have to wait a long time for the answers.

Hipparcos is a bird of an entirely different color. It is not a NASA project but belongs to the European Space Agency, having been invented and originally proposed by Professor Pierre Lacroute of Dijon, France. It is independent of the Shuttle, being small enough to be placed into a geostationary orbit by the French Ariane 1 launch system. And it is cheap enough to be the first of a series. If the first Hipparcos mission works well, it will be easy to launch follow-up missions to give us higher precision or more extensive coverage. If the first Hipparcos fails, it will not be a major disaster.

Hipparcos is an astrometric satellite, designed to do nothing else than measure very accurately the angular positions of stars in the sky. It will give positions about ten times as accurate as those measured by ground-based telescopes. This sounds like a modest and unrevolutionary objective. But in fact the improvement of positional accuracy is of central importance to astronomy. If we can improve the accuracy of angular position measurement by a factor of ten, we increase by a factor of ten the distance out to which we can determine the distances of stars by the method of parallaxes, and so we increase by a factor of a thousand the sample of stars whose distances we can reliably measure. The improved positional accuracy will give us a stereoscopic three-dimensional view of several hundred thousand stars, instead of the few hundred which lie close enough to have parallaxes measurable accurately from the ground. When a star's distance is known, its absolute brightness is also known, and the absolute brightness is the most important quantity determining the structure and life history of a star.

The ten-fold extension of our range of stereoscopic vision will have qualitative as well as quantitative importance. The few hundred stars whose distances we can now measure accurately are a random sample of stars which happen to lie close to the Earth; they are almost all dwarf stars of the commonest types, giving little information about interesting phases of stellar evolution. When we have a sample of several hundred thousand stars of known distance and known absolute brightness, the sample will include many rare types of star, for example, variable stars of various kinds, which we are observing at crucial transient phases of their lives. In this and other ways, the data from Hipparcos will give us a wealth of new information about the constitution and evolution of stars and about the dynamical behavior of our galaxy. The Hipparcos mission includes a completely automated data-processing system on the ground, so that star positions will not be measured laboriously one at a time but will be computed wholesale in batches of several thousand. The data-processing system will be a more revolutionary improvement of the state of the art of astrometry than the satellite itself.

I hope I am not exaggerating the virtues of Hipparcos. I do not wish to embarrass my European friends by giving their brainchild more praise than it deserves. But two facts about Hipparcos seem to me to be of fundamental importance. First, it is the first time since Sputnik in 1957 that a major new development in space science has come from outside the United States. Second, it is the first time since the days of Uhuru that a major new development has come from a small and relatively cheap mission, a mission which could be repeated and further developed without putting excessive strain on launch facilities and budgets. These two facts seem to me to be a good augury for the future. I am inclined to believe that space science will flourish only if we can move away from grand one-shot missions like Hubble Space Telescope and Galileo, toward smaller and more flexible missions in the style of Hipparcos. And Hipparcos is probably only the first of many good ideas which will come from the space-science programs of Europe and Japan, giving us the competitive stimulus which the Soviet space-science program once promised and perhaps will again provide now that Raoul Sagdeev is in charge.

Besides Hipparcos, there are many other opportunities that still lie open for doing first-rate science with smaller missions in space. There are many possible directions in which we may look for a renaissance, a new flowering of space science. Some of these new directions could be started today; others require technology which may take ten or twenty years to develop. I will try not to be too imaginative. I will talk only about possibilities which have some chance of being realized within the next twenty-five years.

I begin with a list of four projects which could be undertaken in the near future. The first of these projects could be starting right now if we had our scientific priorities straight, namely, the Orbiting VLBI Observatory, designed by Bob Preston and his colleagues at the Jet Propulsion Laboratory in California. VLBI means Very Long Baseline Interferometry. During the last twenty-five years the techniques of VLBI have been developed and refined by radio astronomers working with ground-based telescopes in many countries around the world.

VLBI is one of the most spectacular successes in the history of astronomy. The VLBI system already allows us to observe distant radio sources with angular resolution a thousand times better than the best optical telescope on the ground and fifty times better than the best the Hubble Telescope will be able to do. The performance and versatility of ground-based VLBI systems are still improving rapidly and leaving the optical astronomers further and further behind. The secret of success in science is to put your money quickly on the winning horse. The Orbiting VLBI Observatory would be a single radio antenna of modest size, orbiting the Earth in an elongated orbit and adding its signals to the existing network of VLBI telescopes on the ground. The addition of a single space antenna would improve the capabilities of the ground-based system by a factor of about ten. And that would not be the end of the story. Like Hipparcos, the Orbiting VLBI Observatory is a comparatively small cheap mission, small enough to ride piggy-back into orbit with some higher-priority spacecraft, cheap enough to be repeated if it works well. Within ten years we could have a network of orbiting VLBI observatories in a variety of orbits, pushing the angular resolution of radio-astronomical observation toward the ultimate limits set by the lumpiness of the interstellar plasma. And all this on a "learn as you go" basis, free from the rigidities of a one-shot mission.

The second item on my list of good bets for the future is astrometric spacecraft, using and extending the technology of Hipparcos. There is no reason why the United States should leave this area of science entirely to the Europeans. It would be extremely rewarding to extend the range of stereoscopic vision of optical astronomy still further, either by improving the precision of the Hipparcos optical system or by launching Hipparcos spacecraft away from the Earth as far as Saturn to obtain parallaxes on a ten-times-longer baseline. An infra-red version of Hipparcos could look astrometrically at small infra-red sources. That might be even more exciting than the standard Hipparcos. It would see nearby objects with high velocities. It would discover great numbers of asteroids. It would have a good chance of discovering the Nemesis star, the hypothetical companion star of the Sun, if the Nemesis star exists.

The third item on my list is merely to repeat things we have done before. The people in charge of NASA have the idea that a mission once done need never be done again. That is because they are not scientists. If you are an astronomer, you often want to repeat an observation. There are many missions which we would like to do again. They would be cheaper if we did them several times. Why don't we repeat IUE, why don't we repeat IRAS, why don't we repeat the Einstein X-ray satellite? We would learn much more about variable sources if we kept on observing them. Many of the most interesting objects in the sky are variable.

The fourth item on my list is optical interferometry. Interferometry means taking the light or the radiowaves from several receivers and combining it into a single coherent signal before detecting it. The coherent signal from an array of receivers, when fed into a modern data-processing system, allows us to reconstruct an image of the thing we are observing with angular resolution determined by the size of the entire array rather than by the size of individual receivers. Radio interferometers work routinely with arrays many kilometers long, and give angular resolution as good as a single 10-kilometer dish. A dish 10 kilometers in diameter would not only be absurdly expensive but structurally impossible. There is no reason why optical interferometry in space should not be as spectacularly successful as radio interferometry has been on the ground. Optical interferometry does not require large telescopes or large rigid structures. Early missions could be modest, with baselines of a few tens of meters and telescope apertures of a few inches. This would enable us to map the optical structure of bright objects with angular resolution ten times better than the Hubble Telescope. After that, we could develop the technology further so as to reach faint objects and achieve still higher resolution. It took the radio astronomers twenty-five years to learn the art of very-long-baseline interferometry. It will probably take about as long for the practitioners of the art of optical interferometry to catch up with them.

All the suggestions which I have made have been connected with astronomy. This is because astronomy happens to be the science I know best. Similar opportunities exist for new departures in other areas of space science such as solar physics, planetary exploration, and the study of interplanetary particles and fields. I am not claiming that the four items which I listed are necessarily the best or the most important things for a space-science program to do. I am claiming only that these four projects illustrate a new style of operation which could rescue all branches of the space-science enterprise from the doldrums in which they are now stuck.

In astronomy the advantages of the new style are particularly clear. Optical astronomy would move, as radio astronomy has already moved, away from the old technology of single dishes and into the new technology of large arrays, phase-sensitive sensors and sophisticated data processing. The Hubble Telescope which we are scheduled to launch in 1988 is nineteenth-century technology, a big rigid lump of glass. It is the end of the old era rather than the beginning of the new. The new era of astronomy will be flexible, both mechanically and psychologically. If we can begin now to explore in a modest way the technology of flexible optical arrays, we should arrive at the year 2010 with a variety of instruments surpassing the Hubble Telescope in power and versatility as decisively as the radio-astronomical arrays of today surpass the radio telescopes of the 1950s.

The four projects that I mentioned have been somewhat conventional. I now move on to more unconventional things. Now I am talking about things which may be twenty years away. You never know. Sometimes things happen faster than you expect, if you are willing to jump and not wait for some committee to approve them. So I end with four longer-range technological initiatives. The first of these is something that has been talked about at the Jet Propulsion Laboratory in California. It is called the microspacecraft. The enormous advances in data processing during the last twenty-five years have been mainly a result of miniaturization of circuitry. The idea of the microspacecraft is to miniaturize everything, the sensors, the navigational instruments, the communication systems, the antennas and everything else, so that the whole apparatus is reduced in weight like a modern pocket calculator. You still have to have antennas of a certain size, but they can be very thin. The whole thing could be very light. There is no law of physics which says that a high-performance spacecraft like Voyager has to weigh a ton. We might be able to build vehicles to do the same job in the 1-kilogram weight class.

The second of my technological initiatives is a serious effort to exploit solar sails as a cheap and convenient means of transportation around the solar system, at least in the zone of the inner planets and the asteroid belt. The main reason why solar sailing has never seemed practical to the managers of NASA is that the sails required to carry out interesting missions were too big. Roughly speaking, a 1-ton payload requires a square kilometer of sail to drive it, and a square kilometer is an uncomfortably large size for the first experiments in packaging and deploying sails. Nobody wants to be the first astronaut to get tangled up in a square kilometer of sail. But the development of solar sails would be a far more manageable proposition if it went hand in hand with the development of microspacecraft. A 1-kilogram microspacecraft would go nicely with a 30-meter-square sail, and a 30-meter square is a reasonable size to experiment with. The problems of sail management and payload miniaturization will be solved more easily together than separately.

Third on the list of more far-out projects is an idea which I call "minilaser propulsion." I have long been an advocate of laser propulsion. Laser propulsion was invented a long time ago by Arthur Kantrowitz. It is a system of launching rockets from the ground using a laser beam as the energy source, so that the rocket climbs up the laser beam into the sky. The propellant can be something convenient like water. The water is heated up to a high temperature by the laser beam, and it gives such a high exhaust velocity that the rocket can reach orbit or escape from the Earth with a single stage. The classical laser propulsion system of Kantrowitz needed a 1,000-megawatt laser to supply the power. It could launch a payload of 1 ton, which is a typical payload for present-day missions. It needed for that only about 1 ton of propellant, so the vehicle could be small, about the size of a Volkswagen. It would be extremely cheap compared with a conventional three-stage chemical rocket. The trouble was that you had to have your 1,000-megawatt laser first, and that is a very expensive item.

The whole scheme suffers from the grandiose size of the laser. It will take decades to build the laser, and by that time all the interesting science will have changed. However, it might be better if you scale the thing down. I have been looking at a minilaser propulsion system which uses lasers of a few megawatts. It is a comfortable sort of laser; we know how to build it. The motor does not yet exist. Nobody has yet designed a rocket motor that can lift itself off the ground using laser power. It is at the moment about as far along as Robert Goddard's chemical rockets were in the year 1928. But with luck, the motor will work. You could launch little 1-kilogram spacecraft, then, wonderfully cheap; if you calculate how much electricity it takes to get them into orbit, or even to escape from the Earth, it is about a hundred kilowatt-hours per launch. The energy costs are really low. We will need a high volume of traffic in order to make the thing economic. But conceivably, if you had a system like this, it could become cheap enough for ordinary people to use.

The last of my projects is the "space butterfly." That is a way of exploiting for the purposes of space science the biological technology which allows a humble caterpillar to wrap itself up in a chrysalis and emerge three weeks later transformed into a shimmering beauty of legs and antennae and wings. I spoke about butterflies in Chapter 2. Anybody who lives in New Jersey can easily watch a Monarch butterfly climbing into its cocoon and then afterwards climbing out again. It is an awe-inspiring sight. Sooner or later, probably fairly soon, we will understand how that is done. Somehow it is programmed into the DNA and we should soon learn how to do it. It is likely that, within the next twenty-five years, this technology will be fully understood and available for us to copy. So it is reasonable to think of the microspacecraft of the year 2010, not as a structure of metal and glass and silicon, but as a living creature, fed on Earth like a caterpillar, launched into space like a chrysalis, riding a laser beam into orbit, and metamorphosing itself in space like a butterfly. Once it is out there in space, it will sprout wings in the shape of solar sails, thus neatly solving the sail deployment problem. It will grow telescopic eyes to see where it is going, gossamer-fine antennaé for receiving and transmitting radio signals, long springy legs for landing and walking on the smaller asteroids, chemical sensors for tasting the asteroidal minerals and the solar wind, electric-current-generating organs for orienting its wings in the interplanetary magnetic field, and a high-quality brain enabling it to coordinate its activities, navigate to its destination, and report its observations back to Earth.

I do not know whether we will actually have space butterflies by the year 2010, but we shall have something equally new and strange, if only we turn our backs to the past and keep our eyes open for the opportunities which are beckoning us into the twenty-first century.

10

ENGINEERS' DREAMS

There are two ways to predict the progress of technology. One way is economic forecasting, the other way is science fiction. Economic forecasting makes predictions by extrapolating curves of growth from the past into the future. Science fiction makes a wild guess and leaves the judgment of its plausibility to the reader. Economic forecasting is useful for predicting the future up to about ten years ahead. Beyond ten years it rapidly becomes meaningless. Beyond ten years the quantitative changes which the forecast assesses are usually sidetracked or made irrelevant by qualitative changes in the rules of the game. Qualitative changes are produced by human cleverness, the invention of pocket calculators destroying the market for slide rules, or by human stupidity, the mistakes of a few people at Three Mile Island destroying the market for nuclear power stations. Neither cleverness nor stupidity is predictable. For the future beyond ten years ahead, science fiction is a more useful guide than forecasting. But science fiction does not pretend to predict. It tells us only what might happen, not what will happen. It deals in possibilities, not in probabilities. And the most important developments of the future are usually missed both by the forecasters and by the fiction writers. Economic forecasting misses the real future because it has too short a range; fiction misses the future because it has too little imagination.

I took the title of this chapter from one of my favorite books, *Engineers' Dreams* by Willy Ley. Ley's book is concerned with the dreams of the 1930s, the time when Ley was a young man. Ley was a frustrated rocket engineer who later became a successful writer. The dreams which are recorded in his book are mostly projects of civil engineering, enormous dams, tunnels, bridges, artificial lakes and artificial islands. The interesting thing about them is that they are today totally dead. Nobody would want to build them today even if we could afford it. They are too grandiose, too inflexible, too slow. The rules of the political game have changed three or four times in the decades since Willy Ley wrote his book. History passed these dreams by. We do not any longer find it reasonable to think of flooding half of the forests of Zaire in order to provide water for irrigating the deserts of Chad. Instead of huge dams and power stations, soft drinks and machine guns became the hot items in the commerce of Central Africa. Instead of traveling in ships on inland seaways, the people of Africa and South America find it more convenient to use airstrips and light aircraft.

Perhaps it is possible to discern a persistent pattern in the rise and fall of engineering technologies. The pattern resembles in some ways the rise and fall of species in the evolution of plants and animals. A technology during its phase of rapid growth and spectacular success is usually small, quick and agile. As it grows mature it becomes settled and conservative, prevented by the inertia of size from reacting quickly to sudden shocks. When a technology has grown so big and sluggish that it can no longer bend with the winds of change, it is ripe for extinction. Extinction may be long delayed or avoided in sheltered corners, but an overripe technology cannot regain its lost youth. New small and quick alternatives will be waiting to occupy the ecological niche left vacant by the decline of the old. This rhythm, which repeats itself in the evolution of species over periods of millions of years, seems to run about a hundred thousand times faster in the evolution of human technology.

The technology of computers is a particularly clear example of the evolutionary pattern. I am old enough to have been present almost at the creation of computer technology. I was in Princeton in the late 1940s and early 1950s when John von Neumann was building our famous JOHNNIAC computer. Von Neumann was a great mathematician and had the reputation at that time of being the cleverest man in the world. He was supposed to be the intellectual force driving the whole development of computers. He was a great thinker and a great entrepreneur. And yet he totally misjudged the role that computers were to play in human affairs.

I remember a talk that Von Neumann gave at Princeton around 1950, describing the glorious future which he then saw for his computers. Most of the people that he hired for his computer project in the early days were meteorologists. Meteorology was the big thing on his horizon. He said, as soon as we have good computers, we shall be able to divide the phenomena of meteorology cleanly into two categories, the stable and the unstable. The unstable phenomena are those which are upset by small disturbances, the stable phenomena are those which are resilient to small disturbances. He said, as soon as we have some large computers working, the problems of meteorology will be solved. All processes that are stable we shall predict. All processes that are unstable we shall control. He imagined that we needed only to identify the points in space and time at which unstable processes originated, and then a few airplanes carrying smoke generators could fly to those points and introduce the appropriate small disturbances to make the unstable processes flip into the desired directions. A central committee of computer experts and meteorologists would tell the airplanes where to go in order to make sure that no rain would fall on the Fourth of July picnic. This was John von Neumann's dream. This, and the hydrogen bomb, were the main practical benefits which he saw arising from the development of computers.

The meteorologists who came to work with Von Neumann knew better. They did not believe in the dream. They only wanted to understand the weather, not to control it. They had a hard enough time trying to understand it. They tried especially hard to predict a particular hurricane which came up from the Gulf of Mexico and passed close by Princeton in the fall of 1949. Again and again they set the initial conditions in the Gulf of Mexico and tried to predict the hurricane for several years after it happened. So far as I remember, they never did succeed in getting the hurricane on the computer to end up anywhere near Princeton. Now another thirty-five years have gone by, and we have built four more generations of computers, and still we are not doing very well with the prediction of hurricanes. Nobody any longer believes seriously in the possibility of controlling the weather.

What went wrong? Why was Von Neumann's dream such a total failure? The dream was based on a fundamental misunderstanding of the nature of fluid motions. It is not true that we can divide fluid motions cleanly into those that are predictable and those that are controllable. Nature is as usual more imaginative than we are. There is a large class of classical dynamical systems, including non-linear electrical circuits as well as fluids, which easily fall into a mode of behavior that is described by the word "chaotic." A chaotic motion is generally neither predictable nor controllable. It is unpredictable because a small disturbance will produce exponentially growing perturbation of the motion. It is uncontrollable because small disturbances lead only to other chaotic motions and not to any stable and predictable alternative. Von Neumann's mistake was to imagine that every unstable motion could be nudged into a stable motion by small pushes and pulls applied at the right places. The same mistake is still frequently made by economists and social planners, not to mention Marxist historians.

But the misunderstanding of fluid dynamics was not Von Neumann's worst mistake. He not only guessed wrong about the problems to which computers could be usefully applied.

He also guessed wrong about the evolution of the computer itself. Right up to the end of his life, he was thinking of computers as big, expensive and rare, to be cared for by teams of experts and owned by prestigious institutions like Princeton and Los Alamos. He missed totally the real wave of the future which started to roll when outfits like Texas Instruments and Hewlett Packard got into the game. The real wave of the future was to make computers small, cheap and widely available. Nowhere in Von Neumann's writings does one find the shadow of a hint that computers might be sold for a few hundred dollars and be owned by teenagers and housewives. Never did he imagine the flexibility and versatility that computers could achieve once they became small and cheap. He did not dream of the computer-toy industry, the chess-playing machine that fits inside a chessboard and plays a tolerably expert game, the endless varieties of computer-simulated pinball and Dungeons and Dragons. He failed altogether to foresee the rise of the software industry, the buying and selling of computer programs which would help the housewife to organize her recipes and help the teenager to correct the spelling of her homework.

I have a friend, a young American physicist, who spent a year doing theoretical physics in the Soviet Union. He likes to go to the Soviet Union, not because it is a good place to do physics, but because it is a good place to observe the human comedy. When he went back to Leningrad recently for a shorter visit, he received a proposal of marriage and was called in twice for questioning by the KGB, all within the first week. He speaks fluent Russian, and the KGB people find it difficult to believe he is not a spy. He tells me that there is now a flourishing black market in software in the Soviet Union. Designer blue jeans and tape recorders are passé; the new symbols of status among the trendy youth are floppy disks. But there is a shortage of hardware to go with the software. It is not so easy to pick up an IBM PC from a smuggler hanging around a street corner on the Nevskii Prospekt. My friend considers that we in the West are missing a great opportunity to disrupt the economy of the Evil Empire. He says we ought to be flooding the Soviet Union with personal computers and software. This would give a boost to all kinds of private and semi-legal enterprises with which the official state enterprises could not easily compete. The official economy is still living in the Von Neumann era, with big expensive computers under central control. My friend believes that small-computer technology would flow around the apparatus of the state-controlled economy. He finds it more plausible to dream of drowning the Soviet party apparatchiki in a flood of Macintoshes than to dream of starving them into submission with a technological blockade. Small modern computers and software are good tools for eroding the machinery of totalitarian government. That is another engineer's dream which may or may not come true.

I now leave the subject of computers and move to an area with which I am more familiar, the exploration of space. In January 1986 I had the great joy of sitting with the space engineers at the Jet Propulsion Laboratory (JPL) in California while they watched the pictures come in from the Voyager spacecraft encountering Uranus. That was not only a great day for science but a masterpiece of good engineering. The most impressive aspect of the operation was that not everything went smoothly. Several times the engineers had serious problems. Every time there was a problem, they were prepared for it. They had thought ahead. They had Plan B ready as soon as Plan A ran into trouble. At a tense moment at the beginning of the Uranus encounter, an urgent call for help arrived from the Europeans tracking their Giotto spacecraft on its way to Halley's Comet. The Giotto spacecraft had slipped. Its high-gain antenna beam was no longer pointing at Earth. The Europeans had lost contact with it. There was serious danger that the whole Giotto mission would be lost if they could not quickly regain contact. So they telephoned JPL.

JPL runs the Deep Space Net with the three most sensitive receivers in the world, the only receivers capable of picking up the Voyager signals from Uranus. As it happened, the

Deep Space Net dish at Goldstone in California was the only one that could point at the Giotto spacecraft at that moment, and it was also the only one that could point at Uranus. So the JPL engineers changed their schedule so as to hand Voyager over as soon as possible to the Deep Space Net dish in Australia, and then pointed the Goldstone dish at Giotto. Goldstone picked up the feeble signal from Giotto's omnidirectional antenna and transmitted the order to Giotto to reorient the high-gain antenna to the direction of Earth. Within half an hour the high-gain signal from Giotto came in loud and clear, and the control was handed back to Europe. Plan B was immediately put into effect. The Voyager encounter sequence was reprogrammed so as to compensate for the Giotto interruption without losing any scientific data.

The JPL engineers managed to perform another new trick at Uranus which they had not done at Jupiter and Saturn. After passing by Uranus, the spacecraft went behind the planet as seen from the Earth. It was behind Uranus for eighty-three minutes before it reappeared on the other side. During the whole of this hour-long eclipse, the Voyager high-gain antenna was kept accurately pointed at the edge of Uranus, so that the transmitted signal might be refracted around Uranus within the Uranus atmosphere and from there continue on its way to Earth. The signal coming from Voyager to Earth, after creeping around the planet through the Uranus atmosphere, was detected from beginning to end of the eclipse. As a result, we now have precise knowledge of the structure of the Uranus atmosphere down to a pressure of three Earth atmospheres. This at a distance of 2 billion miles with a spacecraft which was launched nine years ago. All of us who were there in the room at JPL were deeply impressed. That was an engineer's dream which came true. We agreed to meet again at JPL on August 25, 1989, for the encounter with Neptune.

Three days after the Uranus encounter, I was flying home to Princeton on a commercial jet with a news program on the television screen and saw the Shuttle blow up. What a sad contrast! The Voyager team at JPL with their superb competence and their triumphant success; the Shuttle, stumbling from one misfortune to another until in the end this last disaster did not come as a surprise. Immediately I thought of the two explorers, the Norwegian Amundsen and the British Scott, who went to the South Pole in the summer of 1911 to 1912. Another of my favorite books is *The Worst Journey in the World,* by Apsley Cherry-Garrard, one of the survivors of the Scott expedition. Cherry-Garrard ends his book with an overall assessment of the expedition in which he had struggled and suffered ten years earlier. Here is his verdict:

> In the broad perspective opened up by ten years' distance, I see not one journey to the pole, but two, in startling contrast one to another. On the one hand, Amundsen going straight there, getting there first, and returning without the loss of a single man, and without having put any greater strain on himself and his men than was all in the day's work of polar exploration. Nothing more businesslike could be imagined. On the other hand, our expedition, running appalling risks, performing prodigies of superhuman endurance, achieving immortal renown, commemorated in august cathedral sermons and by public statues, yet reaching the pole only to find our terrible journey superfluous, and leaving our best men dead on the ice. . . . Any rather conservative whaling captain might have refused to make Scott's experiment with motor transport, ponies and man-hauling, and stuck to the dogs; and it was this quite commonplace choice that sent Amundsen so gaily to the pole and back, with no abnormal strain on men or dogs, and no great hardship either. He never pulled a mile from start to finish.

That is, in a few words, the story of Amundsen and Scott, Amundsen the explorer who knew his business, and Scott the tragic hero who didn't. Almost every word of Cherry-Garrard's verdict applies equally well to Voyager and Shuttle. It is important to remember that there was not one expedition but two. The British in 1913 made Scott into a national hero and ignored Amundsen. The Americans now are making the seven astronauts into national heroes and are ignoring

Voyager. Amundsen's dog-sleds in 1911 took him gaily to the pole, just as the Titan-Centaur launcher in 1977 sent Voyager gaily on its way to Uranus. Nobody said in England in 1913, "Wasn't it stupid of Scott not to use dog-sleds?" and nobody in the official board of enquiry into the shuttle accident said, "Wasn't it stupid of NASA not to use Titan-Centaur?" Most of the payloads which the Shuttle is supposed to carry into orbit could have been launched more conveniently by Titan-Centaur without risking any lives. But NASA decided, after Voyager was on its way, to shut down the production of Titan-Centaur. Scott happened to dislike dogs and NASA happened to dislike Titan-Centaur. Titans occasionally blow up too, but when they blow up it is not a major tragedy.

There are some missions which have human activity in space as a primary purpose. For these the shuttle was an appropriate vehicle. But many missions, and the fatal Challenger mission in particular, carried payloads having nothing to do with human activity. If you want to launch a military or commercial communications satellite, it is much more convenient not to have people on board the launcher. You need to put communications satellites or scientific spacecraft into a variety of orbits, but the exigencies of landing with a human crew restrict the orbit that the Shuttle can reach. When you carry people, you lose flexibility in the choice of orbit as well as in the time of launching.

What should we be learning from the misfortunes of the Shuttle? I hope the American public will not be led to believe that the misfortunes arose merely from poorly designed O-rings. The lessons we ought to be learning are similar to the lessons we have learned from the failure of Von Neumann's dreams of weather control. It was stupid of Von Neumann to push his computers toward one grandiose objective and ignore the tremendous diversity of more modest applications to which computers would naturally adapt themselves. We have learned that the right way to develop a computer industry is to find out first what the customers need and then design the machines to do it. The customers want many kinds of things, so we build many kinds of computers. Most of the customers have modest needs, and therefore most of the computers are small and cheap. These lessons are just as valid in space as they are here on the ground. Customers' needs should drive the industry, not engineers' dreams. The Shuttle was an engineer's dream.

The fundamental mistake of the Shuttle program was the dogmatic insistence of NASA that this single launch system was to take care of all the customers regardless of their needs. Few of the customers wanted the Shuttle. The space-science community hated the Shuttle because it deprived us of the frequent and flexible launch opportunities that science requires. The military hated the Shuttle for similar reasons. The industrial customers, who need the Shuttle mostly for launching communication satellites, hated it less, but they too would have preferred a choice between several launch systems with more flexible schedules. The insistence that Shuttle be the sole launch system was directly responsible for the disaster of January 1986. If alternative launch systems had been available for launching unmanned satellites, there would have been no strong pressure to keep the Shuttle on schedule and no strong reason to fly the Shuttle in bad weather. If the Shuttle had been used only for the missions for which human passengers were essential, it could have waited for good weather and the crew of the Challenger would still be alive and well.

It is interesting to try to imagine what kind of space program we would have today if the program had been driven by customers' needs rather than by engineers' dreams. First of all, we would have kept the passenger business and the freight business separate. The railroads learned long ago that it does not pay to carry passengers on a freight train or to carry coal on a passenger train. If passengers and coal were forced to travel together, the railroads would be in even worse shape than the Shuttle. The launch systems existing in the 1970s, the Scout, Delta and Titan with a variety of upper stages, were well suited to the variety of customers requiring freight transportation in space. We could have satisfied the customers' needs in the freight business merely by keeping these systems in operation. A customer-driven freight business in space would not need any new kinds of launcher. The customers have learned to increase the profitability of space operations by applying new technology to the payloads, not to the launchers. It is far more profitable to use technology to increase the value per pound of the payload rather than to try to decrease the cost per pound of the launch. A customer-driven space program, so far as the freight business is concerned, would still be using the old Scout, Delta and Titan launchers, with the possible addition of a new ultra-light system designed to orbit small payloads as cheaply as possible.

The more difficult and controversial question is, what would a customer-driven manned space program look like? What are the customers' needs for passenger transportation in space? The customers in this case are the astronauts. What kind of a vehicle do the astronauts really need? Here we can again learn a useful lesson from the past. I am so old that I can remember the days when the *Queen Mary* and the *Queen Elizabeth* were carrying passengers in large numbers across the Atlantic and making large profits for the Cunard-White Star steamship company. At that time the two *Queens* were the largest and fastest ships and held the speed record for Atlantic crossings. The British public was proud of its steamships and believed that the *Queens* were profitable because they were the biggest and the fastest ships afloat. But Lord Cunard, the chairman of the steamship line, said that this belief was the opposite of the truth. Lord Cunard said that his steamship line had never intended to build the biggest and fastest ships afloat. The *Queens* were in fact designed to be the smallest and slowest ships that could do what the customers needed, namely, to provide with two ships a reliable weekly service across the Atlantic. They were profitable because they were the slowest and smallest ships that could do the job. The fact that they incidentally broke speed records had nothing to do with it. And the Boeing 707 eventually drove them out of business because it was smaller and could do the job better.

A customer-driven Shuttle would be like the *Queen Mary* and the *Queen Elizabeth.* It would be the smallest and cheapest vehicle that could provide the astronauts with a flexible and reliable service into orbit. It would not need to carry freight. It would not need to carry a crew of seven. It would not need to be heavy. It would not need to strain the limits of rocket and engine technology. On the other hand, it should be in some respects more capable than the existing Shuttle. It should be able to stay in space with its crew for a month. It should possess a larger margin of performance, so that it could reach a greater variety of orbits. And it should be capable of rendezvous and docking with other spacecraft in orbit. A Shuttle with these characteristics is what the astronauts need, and it could probably have been provided more quickly and more cheaply than the existing Shuttle. Above all, what the astronauts need is flexibility. Nobody can predict what they will want to do ten or twenty years ahead. To give them flexibility, the Shuttle should be like a high-performance two-seater airplane, small and light and carrying a big reserve of fuel, capable of jumping to take advantage of unexpected opportunities.

We saw in the operation of the Voyager fly-by at Uranus what flexibility means. The conditions for taking pictures of the moons of Uranus were far less favorable than the conditions which Voyager enjoyed at the encounter with Saturn five years earlier. Sunlight at Uranus is four times dimmer than at Saturn. Longer exposures were required in order to collect the same amount of light. The radio signals received at Earth from Voyager were four times fainter. The instruments on board Voyager were the same at Uranus as they were at Saturn, only five years older. Nevertheless, the pictures of the Uranus moons achieved an angular resolution twice as good as the pictures of the moons of Saturn.

How was this possible? How could the angular resolution be improved by a factor of two without changing a single detail of the hardware? The improvement was possible because the Voyager operation was controlled by on-board software which was almost completely reprogrammable. The software could be, and was, reprogrammed by the JPL engineers at a distance of 2 billion miles. The JPL engineers had five years between Saturn and Uranus to think of improvements and to redesign the software. The new software allowed Voyager to compensate more accurately for the motion of the spacecraft during the taking of the pictures, and the better compensation resulted in better angular resolution. Flexibility means being able to teach an old spacecraft new tricks. When we were sitting at JPL on January 25, 1986, watching the raw unprocessed pictures of Uranus' moon Miranda come in, as sharp and as beautiful as pictures published in *National Geographic,* we knew that flexibility had paid off. It was appropriate that this moon of Uranus happened to be named after the young heroine of Shakespeare's play, *The Tempest.* In the play, Miranda is the one who exclaims: "O brave new world, That has such creatures in it!" And we were sitting there, seeing her Brave New World for the first time.

When we pay our respect to the dead and build our public monuments to the crew of the Challenger, we should not forget Voyager and the engineers who built her. It is good that we give honor to the brave victims of our stupidity, just as we gave honor to Robert Scott. But we should also give honor to those who were clever and survived, to Roald Amundsen and his dog team, to the JPL engineers who used their brains and reprogrammed the Voyager software 2 billion miles away to give us our first glimpse of a new world.

I now descend from the sublime to the practical, from Miranda to the grubby problems of the electric utility industry here on this planet. I will talk a little about power stations and gas turbines. In the electric utility business we have made mistakes which are similar to the mistakes we made in the space business and in the early days of the computer business. The chief mistake is to believe that the customers will demand something merely because we wish to supply it. Von Neumann believed that his customers would demand weather control and hydrogen bombs, but the customers preferred to spend their money on Pacman. In the space business, the customers preferred expendable launchers and we insisted on supplying them with the Shuttle. In the electric utility business, the suppliers built a number of very large and expensive central power stations which became white elephants because the demand for electricity fell far short of the suppliers' estimates. For the last ten years, customer demand has grown at about half the rate predicted by the suppliers. And nobody can tell whether the demand will continue to shrink or will start growing rapidly again during the next ten years. Predicting electricity demand is almost as hazardous an occupation as predicting the stock market. In the past, the predictions of demand were always too high. In the future, because of the natural human tendency to overreact to disasters, the predictions might easily be too low. Either way, the electric utility industry is in trouble.

I am indebted to my friend Bob Williams at the Princeton Center for Energy and Environmental Studies for my education in the problems of the utility business. The dilemma of the utilities arises from a collision between two facts. First, nobody can predict demand ten years ahead. Second, it takes about ten years to build and bring into operation a large modern power station, whether it is coal-fired or nuclear. The special problems of nuclear power stations are more notorious, but the financial and environmental miseries involved in building a coal-fired station are almost as bad. The ten-year construction time imposes enormous burdens on the utility industry in two ways. It brings a major escalation of costs through interest charges and inflation, and it makes it impossible to match supply to demand. Why do power stations take so long to build? Mostly because the individual units are very large, and the builders build them large in pursuit of economies of scale. Steam generators and steam turbines are more economical in large sizes. For this reason, both nuclear and coal-fired stations were driven to large sizes. Large sizes, large unit costs, and inflexible schedules have brought the utility business to an impasse almost as severe as the impasse facing the Shuttle.

Fortunately, a new technology is now available which offers to the utility industry an escape from the impasse. The new technology is called STIG, for Steam-Injected Gas Turbines. An even better version of STIG called ISTIG (for Intercooled Steam-Injected Gas Turbines) will be available in three or four years. The basic reason why STIG technology is good is that it uses the results of forty years of intensive development of gas turbines by the aircraft engine industry. Steam injection and intercooling convert an aircraft jet engine into a cheap and efficient producer of power. STIG power units are already working profitably as co-generators of electricity and process steam. The fundamental advantage of STIG for the utility industry is that the power units are small and can be installed in two or three years. If supply can follow demand with a lag of only two or three years, the industry's worst problem is solved. The STIG units are about as efficient in small sizes as in large. They do not have important economies of scale. Therefore they need never be larger than required by the local demand. They happen fortuitously to have other essential virtues. They are, as a result of the demanding requirements of the aircraft industry in which they originated, quick to replace and easy to repair. They are, because of the effect of steam on flame chemistry, inherently low in nitrogen-oxide emissions. They comply easily with environmental emission regulations.

Unfortunately, the STIG technology has one major defect to balance its virtues. It is illegal in the United States, except for co-generation units. STIG uses natural gas as fuel, and the Power Plant and Industrial Fuel Use Act of 1978 decrees that no new central power station may burn natural gas. Unless the act of 1978 is repealed or amended, the vendors of STIG in the United States will be producing for export only. Fortunately, the rest of the world is not so encumbered with legislation. In many countries, including developing countries which are now critically dependent on imported oil, abundant resources of natural gas are available. For these countries, STIG will be a godsend. It will enable them to produce electricity cheaply and in conveniently small units. Even in the United States, the resources of natural gas would be large enough to permit STIG units to operate for a thirty-year lifetime if this were permitted by law. One may hope that the 1978 act may be amended, at least to the extent of allowing the replacement of existing inefficient gas-burning units by STIG units burning less gas and producing more electricity.

But we must be careful not to push STIG too hard. We must not assume that, just because engineers like STIG, the customers are bound to like it too. Let us make it available at a reasonable price and then see what happens. Like other engineers' dreams in the past, STIG may not be what the customers want.

So far in this chapter I have been playing the critic and the skeptic. I have been describing engineers' dreams which failed, and emphasizing the folly of the dreamers. Now in the last few pages I switch roles and play the dreamer myself. My dream is science fiction and not science. Its purpose is to enlarge our imagination, to give us a glimpse of a future that we cannot hope to predict in detail.

I am dreaming of the next space mission to explore the planet Uranus. I call the mission Uranus 2, and I imagine it arriving at Uranus in the year 2016. The Voyager fly-by was only a beginning. Like all good scientific missions, it raised more new questions than it answered old ones. A single fly-by is not enough. We need to explore Uranus comprehensively and thoroughly, to study its chemistry and meteorology and evolutionary history, to study the magnetic field and the rings, the topography and geology of each of the moons.

This is an ambitious program. It is unlikely that it can be accomplished by further missions in the style of Voyager. The Voyager fly-by comes close to the limit of what can be achieved by the technology of the 1970s. I am assuming that when we go back to Uranus next time, we will go with the technology of the twenty-first century. From Voyager to Uranus 2 is a big jump, as big a jump as we made from the primitive sounding rockets of the 1940s to Voyager. From the

V-2 to Voyager was thirty years, and from Voyager to Uranus 2 will be another thirty years. It would have been difficult in 1946 to imagine Voyager, and it is difficult today to imagine Uranus 2. The difficulty in imagining the future comes from the fact that the important changes are not quantitative. The important changes are qualitative, not bigger and better rockets but new styles of architecture, new rules by which the game of exploration is played. It would have been difficult to imagine Voyager in 1946 because the concept of on-board reprogrammable software did not then exist. It is difficult to imagine Uranus 2 today because the concept of a biologically organized spacecraft does not yet exist.

The Uranus 2 mission is a good one to dream about. It is too far away to be reached by our existing technology, but not too far away to be achieved by young people alive today if they are not afraid to break new paths into the future. I am not saying that my version of the Uranus 2 mission will happen, or that it ought to happen. The real future of space science may turn out to be quite different. All I am saying is that my version of Uranus 2 is a possibility, and that it is not too soon to begin thinking about it. Uranus 2 will not be a one-shot mission like Voyager. If Uranus 2 flies as I imagine it, it will be only one among a large flock of similar birds flying out to various destinations all over the solar system. Before Uranus 2 flies, her cousins will already be exploring Mars and Jupiter and Saturn. Uranus 2 will be cheap enough to fly frequently. The spacecraft that flies the Uranus 2 mission will be called Astrochicken. I want it to have a name like itself, cheerful and unpretentious.

The basic idea of Astrochicken is that the spacecraft will be small and quick. I do not believe that a fruitful future for space science lies along the path we are now following, with space missions growing larger and larger and fewer and fewer and slower and slower as the decades go by. I propose a radical step in the direction of smallness and quickness. Astrochicken will weigh a kilogram instead of Voyager's ton, and it will travel from Earth into orbit around Uranus in two years instead of Voyager's nine. The spacecraft must be far more versatile than Voyager. It must land on each of Uranus' moons, roam around on their surfaces, see where it is going, taste the stuff it is walking on, take off into space again, and navigate around Uranus until it decides to make a landing somewhere else. To do all this with a 1-kilogram spacecraft sounds crazy to people who have to work and plan within the constraints of today's technology. Perhaps it will still be crazy in 2016. Perhaps not. I am dreaming of the new technologies which might make such a crazy mission possible.

Three kinds of new technology are needed. All three are likely to become available for use by the year 2016. All three are already here in embryonic form and are advanced far enough to have names. Their names are genetic engineering, artificial intelligence and solar-electric propulsion. Genetic engineering is fundamental. It is the essential tool required in order to design a 1-kilogram spacecraft with the capabilities of Voyager. Astrochicken will not be built, it will be grown. It will be organized biologically and its blueprints will be written in the convenient digital language of DNA. It will be a symbiosis of plant and animal and electronic components. The plant component has to provide a basic life-support system using closed-cycle biochemistry with sunlight as the energy source. The animal component has to provide sensors and nerves and muscles with which it can observe and orient itself and navigate to its destination. The electronic component has to receive instructions from Earth and transmit back the results of its observations. During the next thirty years we will be gaining experience in the art of designing biological systems of this sort. We will be learning how to coordinate the three components so that they work smoothly together.

Artificial intelligence is the tool required to integrate the animal and electronic components into a working symbiosis. If the integration is successful, Astrochicken could be as agile as a hummingbird with a brain weighing no more than a gram. The information-handling apparatus is partly neural and partly electronic. An artificial intelligence machine is a computer designed to function like a brain. A computer of this sort will be made compatible with a living nervous system, so that information will flow freely in both directions across the interface between neural and electronic circuits.

The third new technology required for Uranus 2 is solar-electric propulsion. To get from Earth to Uranus in two years requires a speed of 50 kilometers per second, too fast for any reasonable multistage chemical rocket. It is also too fast for solar sails. Nuclear propulsion of any kind is impossible in a 1-kilogram spacecraft. Solar-electric propulsion is the unique system which can economically give a high velocity to a small payload. In this system, solar energy is collected by a large, thin antenna and converted with modest efficiency into thrust. The spacecraft carries a small ion-jet motor which uses propellant sparingly and gives an acceleration of the order of a milligee.

Nobody has yet done the careful engineering development to demonstrate that the energy of sunlight can be converted into thrust with a power-to-weight ratio of 1 kilowatt per kilogram. That is what Uranus 2 needs. But solar-electric propulsion is probably an easier technology to develop than genetic engineering and artificial intelligence. Since I am talking science fiction, I shall assume that all three technologies will be available for our use in 2016. I can then give a rough sketch of the Uranus 2 mission.

The mission begins with a conventional launch taking the spacecraft from Earth into orbit. Since the spacecraft weighs only 1 kilogram, it can easily ride on any convenient launcher. During the launch, the spacecraft is packaged into a compact shape, and the biological components are busy reorganizing themselves for life in space. During this phase the spacecraft is a fertilized egg, externally inert but internally alive, waiting for the right moment to emerge in the shape of an Astrochicken. After it is in a low Earth orbit, it will emerge from its package and deploy the life-support apparatus needed for survival in space. It will deploy, or grow, a thin-film solar collector. The collector weighs 100 grams and collects sunlight from an area of 100 square meters. It feeds a kilowatt of power into the little ion-drive engine which sends the spacecraft on its way with a milligee acceleration sustained for several months. This is enough to escape from Earth's gravity and arrive at Uranus within two years. The same 100-square-meter collector serves as a radio antenna for two-way communication with Earth. This is ten times the area of the Voyager high-gain antenna. For the same rate of information transmitted, the transmitter power of Astrochicken can be ten times smaller than Voyager, 2 watts instead of 20 watts.

The spacecraft arrives at Uranus at 50 kilometers per second and grazes the outer fringe of the Uranus atmosphere. The 100-square-meter solar collector now acts as an efficient atmospheric brake. Because the collector is so light, it is not heated to extreme temperatures as it decelerates. The peak temperature turns out to be about 800 Celsius or 1500 Fahrenheit. The atmospheric braking lasts for about half a minute and produces a peak deceleration of 100 gees. The spacecraft leaves Uranus with speed reduced to 20 kilometers per second and passes near enough to one of the moons to avoid hitting Uranus again. It is then free to navigate around at leisure among the moons and rings. The solar-electric propulsion system, using the feeble sunlight at Uranus, is still able to give the spacecraft an acceleration of a tenth of a milligee, enough to explore the whole Uranus system over a period of a few years.

The spacecraft must now make use of its biological functions to refuel itself. First it navigates to one of the rings and browses there, eating ice and hydrocarbons and replenishing its supply of propellant. If one ring tastes bad it can try another, moving around until it finds a supply of nutrients with the right chemistry for its needs. After eating its fill, it will use its internal metabolic processes with the input of energy from sunlight to convert the food into chemical fuels. Chemical fuels are needed for jumping onto moons and off again. Solar-electric propulsion gives too small a thrust for jumping. The spacecraft carries a small auxiliary chemical rocket system for this purpose. We know that a chemical rocket system is biologically possible, because there exists on the Earth a creature called the Bombardier beetle which uses a chemical rocket to bombard its enemies with a scalding jet of hot liquid. It manufactures chemical fuels within its body and combines them in its rocket chamber to produce the scalding jet. Astrochicken will borrow its chemical rocket system from the Bombardier beetle. The Bombardier beetle system will give it the ability to accelerate with short bursts of high thrust to escape from the feeble gravity of the Uranus moons. The spacecraft may also prefer to use the Bombardier beetle system for jumping quickly from one place to another on a moon rather than walking laboriously over the surface. While living on the surface of a moon, the Astrochicken will continue to eat and to keep the Bombardier beetle fuel tanks filled. From time to time it will transmit messages to Earth informing us about its adventures and discoveries.

That is not the end of my dream, but it is the end of my chapter. I have told enough about the Uranus 2 mission to give the flavor of it. The underlying idea of Uranus 2 is that we should apply to the development of technology the lessons which nature teaches us in the history of the evolution of life. Birds and dinosaurs were cousins, but birds were small and agile while dinosaurs were big and clumsy. Big main-frame computers, nuclear power stations and Space Shuttle are dinosaurs. Microcomputers, STIG gas turbines, Voyager and Astrochicken are birds. The future belongs to the birds. The JPL engineers now have their dreams on board the Voyager speeding on its way to Neptune. I hope the next generation of engineers will have their dreams riding on Uranus 2 in 2016.

11

THE BALANCE OF POWER

"There is only one path to peace and security: the path of supranational organization. One-sided armament on a national basis only heightens the general uncertainty and confusion without being an effective protection."

That is the voice of Albert Einstein, speaking as always simply and clearly. He spoke these words in 1948, when the United Nations had failed to establish an international control of nuclear activities and the nuclear arms race was beginning. Einstein deduced from that situation what he considered to be the inescapable consequence. National sovereignty, he said, was a luxury mankind could no longer afford.

I use Einstein's words as the text for my sermon. The general subject of this and the next four chapters will be the political problems of arms control and defense. The state of the world today is not essentially different from the state of the world in 1948. We are still faced with the same choices that we were facing in 1948. On the level of fundamental principles, we are faced with a choice between unity and diversity. The unity of mankind, or the diversity of nations and political institutions. National sovereignty is the contemporary expression of the ancient human tradition which divided us into tribes, each jealously guarding its independence and proud of its differences from its neighbors. The diversity of tribes and cultures is, whether we like it or not, deeply rooted in our history. War is the price we have paid for this diversity. It is a price which mankind has usually been willing to pay, whenever the means of warfare were available and the prospects of success not altogether hopeless. Even the tribe of Israel, which made an exception to the general rule by maintaining its separate existence without an army and without war for two thousand years, has in modern times chosen national sovereignty as a preferable alternative, accepting the attendant burdens of intermittent warfare. Einstein, as a member of the tribe, understood and reluctantly supported its violent assertion of national sovereignty. While the war of Israeli independence was still raging in 1948, he wrote: "I never considered the idea of a state a good one, for economic, political and military reasons. But now, there is no going back, and one has to fight it out."

Einstein, in spite of his lifelong opposition to nationalism, had to applaud the Jewish victory of 1949. "What has been achieved," he wrote after the 1949 armistice agreement between Israel and Egypt, "can only be admired." But his admiration for Jewish military prowess was never wholehearted. He continued to believe that the pursuit of national security through national armament was a dangerous delusion. Standing resolutely against the rising tide of nationalism, he continued to preach the necessity of a supranational government. He continued to insist that, in a world armed with nuclear weapons, the only way to avoid total destruction was the federation of mankind.

Einstein is not the only historical figure who has taken a stand for political unity and against tribal diversity. On many occasions in past centuries the voices of unity have prevailed. Here is Queen Anne of England, in her letter of July first 1706 to the Scottish Parliament, arguing the merits of a political union with England.

An entire and perfect union will be the solid foundation of lasting peace. It will secure your religion, liberty and property; remove the animosities amongst yourselves, and the jealousies and differences between our two kingdoms. It must increase your strength, riches and trade. And by this union the whole island, being joined in affection and free from all apprehensions of different interest, will be enabled to resist all its enemies. We most earnestly recommend to you calmness and unanimity in this great and weighty affair, that the union may be brought to a happy conclusion, being the only effectual way to secure our present and future happiness, and disappoint the designs of our and your enemies, who will doubtless, on this occasion, use their utmost endeavours to prevent or delay this union.

The Queen had her way. Six months later, the Scottish Parliament voted by a three-fifths majority to put an end to its independence and to accept the Union. The Queen had been careful to sweeten the deal with a generous cash payment which nominally compensated Scotland for assuming a share of the national debt of England. I leave it to the Scots to decide whether the history of Scotland over the subsequent three centuries has proved the wisdom of the Scottish Parliament's decision to surrender.

After Queen Anne, and quoting her in support of their own arguments, came the American Federalists. Hamilton, Madison and Jay, the authors of the *Federalist Papers* of 1787–88, set down on paper the classic statement of the case for unification. They were writing to persuade the voters of New York State to ratify the Constitution of the United States of America. The ratification of the Constitution was a formal surrender of the sovereignty of New York State, similar to the surrender of the sovereignty of Scotland eighty years earlier. The Federalists convinced the voters, and the Constitution was duly ratified. The *Federalist Papers,* though written in haste and in the context of a local political battle, gave a lasting intellectual legitimacy to the idea of federal government. They still speak to mankind today with undiminished force, warning us of the disasters to which the competition of local sovereignties can lead us.

Here is Alexander Hamilton in the sixth *Federalist Paper:*

> A man must be far gone in Utopian speculations who can seriously doubt that, if these States should either be wholly disunited, or only united in partial confederacies, the subdivisions into which they might be thrown would have frequent and violent contests with each other. To presume a want of motives for such contests as an argument against their existence, would be to forget that men are ambitious, vindictive and rapacious. To look for a continuation of harmony between a number of independent, unconnected sovereignties in the same neighbourhood, would be to disregard the uniform course of human events, and to set at defiance the accumulated experience of ages.

> The causes of hostility among nations are innumerable. There are some which have a general and almost constant operation upon the collective bodies of society. Of this description are the love of power or the desire of preeminence and domination—the jealousy of power, or the desire of equality and safety. There are others which have a more circumscribed though an equally operative influence within their spheres. Such are the rivalships and competitions of commerce between commercial nations. And there are others, not less numerous than either of the former, which take their origin entirely in private passions; in the attachments, enmities, interests, hopes and fears of leading individuals in the communities of which they are members. Men of this class, whether the favorites of a king or of a people, have in too many instances abused the confidence they possessed; and assuming the pretext of some public motive, have not scrupled to sacrifice the national tranquillity to personal advantage or personal gratification. . . .

> Have republics in practice been less addicted to war than monarchies? Are not the former administered by *men* as well as the latter? Are there not aversions, predilections, rivalships, and desires of unjust acquisitions, that affect nations as well as kings? Are not popular assemblies frequently

subject to the impulses of rage, resentment, jealousy, ava-rice, and of other irregular and violent propensities? Is it not well known that their determinations are often governed by a few individuals in whom they place confidence, and are, of course, liable to be tinctured by the passions and views of those individuals? Has commerce hitherto done anything more than change the object of war? Is not the love of wealth as domineering and enterprising a passion as that of power or glory? Have there not been as many wars founded upon commercial motives since that has become the prevail-ing system of nations, as were before occasioned by the cupidity of territory or dominion? Has not the spirit of commerce, in many instances, administered new incentives to the appetite, both for the one and for the other? Let experience, the least fallible guide of human opinions, be appealed to for an answer to these inquiries. . . .

Is it not time to awake from the deceitful dream of a golden age, and to adopt as a practical maxim for the direc-tion of our political conduct that we, as well as the other inhabitants of the globe, are yet remote from the happy empire of perfect wisdom and perfect virtue?

Hamilton is not only a great master of English prose; he is also a great statesman and political thinker. He played a major part in the writing of the American Constitution as well as in selling it to the voters of New York State. The practical success and durability of the Constitution owe much to Hamil-ton's jaundiced view of human nature. The American Consti-tution is designed to be operated by crooks, just as the British constitution is designed to be operated by gentlemen. Because Hamilton believed that men are by nature crooks rather than gentlemen, he was able to help design a constitution that could deal effectively with President Nixon. If ever a World Gov-ernment should come into existence, it had better be a govern-ment designed to be run by crooks rather than a government designed to be run by gentlemen. Gentlemen are too often in short supply.

It would be difficult to find two individuals more opposite in character than Hamilton and Einstein. Hamilton, soldier and politician, arrogant, self-assertive and quarrelsome; Einstein, scientist and philosopher, modest, meditative and quiet. Nevertheless, in its main thrust, Hamilton's argument was the same as Einstein's. Einstein believed as strongly as Hamilton in the darkness at the core of human nature. In a letter to Sigmund Freud in 1932, Einstein wrote, "Man has within him a lust for hatred and destruction. In normal times this passion exists in a latent state, it emerges only in unusual circumstances; but it is a comparatively easy task to call it into play and raise it to the power of a collective psychosis." Einstein agreed also with Hamilton's prescription for taming the human animal. Hamilton was dealing with North America and Einstein with the world as a whole, but the prescription in both cases was the same: a unified legal authority holding a monopoly of military power.

The case for World Government as Hamilton and Einstein presented it is logically compelling. Fortunately or unfortunately, World Government has a fatal defect. Nobody wants it. At the time when Einstein preached World Government most seriously, he was bitterly attacked by Soviet scientific colleagues as well as by patriotic American politicians. "By the irony of fate," the Soviet scientists proclaimed, "Einstein has virtually become a supporter of the schemes and ambitions of the bitterest foes of peace and international cooperation." At the same time the conservative American press was portraying Einstein as a dupe of Soviet propaganda. The idea of World Government had a unique ability to unite the patriots of all countries against it. It was simultaneously seen as a capitalist plot by the Communists and as a Communist plot by the capitalists.

In the thirty years since Einstein died, the prospects for World Government have not improved. On the contrary, the liberation of the colonial empires and the emergence of a hundred new independent countries has made the spirit of nationalism even stronger and more pervasive than before. The new countries, having in many cases achieved their independence by hard fighting, are not likely to surrender it to a supranational authority in which the old colonial masters play a major role. And the old countries of Europe are not likely to surrender their sovereignty to an authority in which they are permanently outvoted by Asians and Africans. The disparities of interest and culture between North and South are an even greater obstacle to World Government than the disparities between Communists and capitalists.

So World Government is not a road to peace, at least for the next few centuries. The human species values its diversity too highly. If we are to find a practical road to peace, we must not try to abolish nationalism. We must try instead to make nationalism an ally in the search for peace. This means that we have to create an international order in which national pride and national power are put to constructive use. National power must be the instrument by which stability of the international order is achieved. We are then driven back to the old concept of a balance of power. National power must be balanced in a stable equilibrium. Nations must be strong enough to defend their own territory but not strong enough to be tempted by dreams of conquest. The historic problem of international diplomacy has been to create a political framework within which a stable balance of power could be maintained. Over the centuries, this problem has rarely been solved. The balance of power has been hard to achieve, and harder still to maintain over long periods of time. Europe has been the area of the world in which the balance of power has been most assiduously pursued, and Europe has also been for centuries the scene of the greatest and most destructive wars. The history of Europe proves conclusively that the balance of power will not be stable if the behavior of nations continues to follow the patterns of the past. The patterns of the past include such characters as Napoleon, Bismarck and Hitler. Alexander Hamilton would not have been surprised by the appearance of such characters. He told us, before they appeared to disturb the peace of Europe, that men are ambitious, vindictive and rapacious. If we are to avoid great wars in the future, the patterns of our behavior must change.

The invention of nuclear weapons might perhaps be a shock sharp enough to change the patterns. That is the hope which underlies the grotesque deployments of nuclear weapons around the world. The hope is that nuclear weapons, deployed in huge numbers and in great variety, will finally make the balance of power stable. The hope is that even a Napoleon or a Hitler will understand, when his country is a target for 10,000 nuclear warheads, that conquest in the grand style is no longer possible. This hope is justified in so far as nuclear weapons have made a repetition of World War II impossible. Never again can a great army walk overland to Moscow. Its homeland will be utterly destroyed before it reaches Smolensk. But unfortunately, not all great wars were deliberately launched as World War II was launched by Hitler. World War I is the prime example of a great war which grew out of muddle and miscalculation. The emperors Franz Josef of Austria and Wilhelm of Germany were not planning a war of conquest in the style of Hitler. They muddled their way into a great war because they did not know how to keep a small war from spreading. World War I showed that the balance of power may be unstable even when there is no Hitler deliberately planning to upset it. It is not at all clear that nuclear weapons have eliminated the danger of a sequence of miscalculations such as occurred in 1914. The risk of a great war arising out of some local squabble by muddle and miscalculation still exists. All that we know for certain is that nuclear weapons vastly increase the cost of such miscalculations. It is wrong to claim that nuclear weapons have made the balance of power reliably stable. We do not know how reliable the existing precarious stability may be. And that is why the perpetuation of nuclear deployments is morally unacceptable.

If World Government is unfeasible and the present system of nuclear-armed national sovereignties is unacceptable, what alternative possibilities are open to us? This is the question which was addressed in a forthright fashion by the Catholic Bishops of the United States in their Pastoral Letter of 1983, "The Challenge of Peace: God's Promise and Our Response."

The bishops begin their discussion where we all ought to begin, with fundamental principles of morality. Nuclear weapons are unacceptable as a basis for a permanent system of international security, not only because they expose us to unacceptable dangers, but even more because they express our intention under certain conditions to commit unacceptable crimes. In particular, the bishops condemn three aspects of our present nuclear deployments: (1) deliberate targeting of civilian populations; (2) willingness to use nuclear weapons first under some circumstances; and (3) planning to fight limited nuclear wars. Items 1 and 2 are condemned as inconsistent with the proportionality of means to ends which is an essential requirement of a just war according to Christian doctrine. Item 3 is condemned on practical grounds, because the bishops were convinced that a limited nuclear war is unlikely to remain limited.

The bishops then come to the basic question of nuclear policy. If these three prohibitions are taken seriously, what is there left for nuclear weapons to do? The bishops are careful not to advocate unilateral nuclear disarmament. They recognize a moral value in the possession of nuclear weapons for the purpose of preventing war. But the permissible retention of nuclear weapons is subject to two severe limitations. First, the weapons should be intended for use, if at all, only in retaliation after a nuclear attack. Second, the possession of the weapons must be a temporary expedient, coupled with serious efforts to achieve multilateral nuclear disarmament by negotiation. The reliance on nuclear weapons is acceptable only if it is not regarded as a permanent solution of our security problem. The long-range goal must be to do away with nuclear weapons altogether.

I happen to agree with the bishops' diagnosis of our situation. It says that when we look to the long-term survival of our civilization, there is only one practical alternative to World Government. The practical alternative is a gradual transition to a non-nuclear world. The non-nuclear world is a world of sovereign nations, each prepared to defend its territory and its vital interests with effective non-nuclear forces, and all committed by international agreement to keep nuclear weapons outlawed. Nobody can guarantee that a non-nuclear world will be either stable, peaceful or safe. But it is at least possible that it will be substantially less dangerous than the world we are living in today. One thing can be said with certainty. The non-nuclear world will not be building its security upon a defiance of our deepest ethical principles.

The next question is, How can we move from the world we live in to a non-nuclear world? How can we imagine getting from here to there? Many of my friends consider me to be, in Alexander Hamilton's words, a man far gone in Utopian speculations. They agree with me that a non-nuclear world would be desirable, but they consider it a Utopia and they disbelieve in the possibility of the transition. In spite of my friends and in spite of Alexander Hamilton, I continue to believe that the non-nuclear world is an attainable goal.

The next three chapters will discuss practical measures which may help to make the transition to a non-nuclear world possible. Two essential tools are required for the transition. The tools are nuclear arms control and non-nuclear defense. Nuclear arms control is obviously required, and it must be arms control of a far more drastic kind than we have recently contemplated. But a sharp improvement in the capabilities of non-nuclear defense is equally necessary. For the transition to a non-nuclear world to be credible, governments must be persuaded that they can defend themselves better without nuclear weapons than with them. Every improvement in non-nuclear defense will make this persuasion easier. Indeed, it is remarkable how willing the majority of governments have been to leave nuclear weapons aside. Nuclear weapons are not particularly expensive or difficult to make. Twenty or thirty of the world's governments could easily produce and deploy nuclear weapons if they saw any clear advantage in doing so. The fact that they have not done so is evidence that nuclear weapons are widely perceived as causing more headaches than they are worth. Presumably, the governments which could

have built nuclear weapons but chose not to do so have calculated that non-nuclear defense gives them the best chance
of staying out of trouble.

Non-nuclear defense comprises two separate missions
with very different status. On the one hand there is tactical
non-nuclear defense, with the mission of defending a country
against non-nuclear attack or invasion. This is the traditional
mission of national defense, and it is generally taken care of
by traditional means such as soldiers, guns, fortifications, airplanes and ships. On the other hand there is strategic non-
nuclear defense, with the mission of defending a country
against nuclear attack. Many people consider strategic defense
of any kind to be a dangerous illusion. The orthodox view
among my academic colleagues is that strategic non-nuclear
defense is unfeasible, whether or not it is desirable. My own
view is that it is desirable, whether or not it is feasible. The
subject of strategic non-nuclear defense came to public attention as a result of President Reagan's Strategic Defense Initiative, otherwise known as the "Star Wars" program. The Star
Wars program as it has been presented to the public is a
strange mixture of fact and fantasy. In Chapter 12 I will try to
disentangle the fact from the fantasy.

The intensity of public discussion of the Star Wars program has had the unfortunate result of giving people an exaggerated idea of its importance. The program has been
oversold by its friends and overrated by its enemies. It is far
less important than the conventional apparatus of armies and
navies with which countries are accustomed to defend themselves. The stability of international relationships depends
more on political than on military factors, and depends more
on conventional than on unconventional armaments. The essential requirement to enable a country to withstand threats to
its independence is political cohesion. If political cohesion is
lacking, exotic weapons will not help. If political cohesion
exists and is backed up by an adequate army, exotic weapons
will probably not be needed. Nevertheless, the Star Wars
program has become a symbol upon which intense hopes and

fears are focused. The hopes and fears are real, even if the program itself is illusory. I therefore set aside a chapter to describe the present status of the program. My description will be based in part on inside information, but it is not in any sense an official statement. My official connection with the program has been as a tame critic and not as a participant.

12

STAR WARS

Strategic defense suddenly became a hot political issue in the United States when President Reagan launched the Strategic Defense Initiative in his speech of March 23, 1983. I have to confess that this speech caused me considerable embarrassment. Like most of my friends in the academic world, I had grown accustomed to being in opposition to President Reagan's policies. Especially in the area of foreign policy, I have found most of President Reagan's statements to be insensitive and ill-informed. It came as a shock on March 23, 1983, to hear the President saying something sensible. It was not only surprising but embarrassing to find myself agreeing with what he said. Here are the three sentences which contain the essence of his proposal:

"Tonight, consistent with our obligations of the ABM Treaty and recognizing the need for closer consultation with our allies, I'm taking an important first step. I am directing a comprehensive and intensive effort to define a long-term research and development program to begin to achieve our ultimate goal of eliminating the threat posed by strategic nuclear missiles. This could pave the way for arms control measures to eliminate the weapons themselves."

I agreed with what he said because I have for a long time believed that a shift in emphasis from offensive to defensive weaponry is likely to improve our chances of establishing a stable world. I start from a fundamental moral prejudice which prefers weapons of self-defense to weapons of mass destruction, and I heard this same prejudice resounding in the President's words. "Wouldn't it be better to save lives than to avenge them?" he said. The driving force behind the President's initiative was his personal feeling of outrage when contemplating the possibility that he might one day find himself responsible for ordering the death of millions in nuclear retaliation after a Soviet attack. This feeling of outrage is widely shared among those who have responsibility for commanding nuclear forces. It ought to be shared by all of us who live under the dubious protection of the nuclear umbrella.

Unfortunately, the basic ethical concerns which underlay the President's initiative were rapidly obscured in the subsequent public debate. The President's speech was dubbed "the Star Wars speech," and the discussion of it concentrated upon technical details of various exotic forms of weaponry in space. The media portrayed the defense initiative as a program for deploying in space giant laser battle stations and other similar absurdities. It was not difficult for opponents of the initiative to prove that laser battle stations would be militarily ineffective. In fact the President in his speech did not once use the word "space," nor did he mention high-powered lasers or other death-ray devices associated in the public mind with the phrase "Star Wars." He merely proposed a program of research and development to explore the possibilities of strategic defense, to find out what works and what doesn't work. If, as seems likely, things on the ground work better than things in space, the end result of the program will bear little resemblance to the technology we saw in the *Star Wars* movie. My own guess is that a militarily effective defense will be largely ground-based, decentralized, widely proliferated and non-nuclear. But it is foolish to speculate about the technical outcome of a long-range research program which is only just beginning. If we are to form a rational judgment of the defense initiative, we should be debating its ethical and political goals and not its technical details. Ends are more important than means. I hope that the public discussion of strategic defense may some day escape from our obsession with technical means and come back to the human ends which the technology is supposed to serve.

I wish now to examine the Star Wars program briefly on its merits, to see what is wrong with it and what is right with it. If we can see clearly what is wrong with it, we may understand better what is right. Broadly speaking, three things are wrong with the Star Wars initiative. First, a large fraction of it, particularly the part concerned with huge optical systems in space, is technical nonsense. Second, another large fraction of it, particularly the part concerned with X-ray lasers, is military nonsense. And third, the program is hidden in such a thick fog of secrecy that neither the American public nor the Soviet government can separate the sense from the nonsense. If I were running the program, the first thing I would do would be to declassify the whole thing. The benefits of declassification would enormously outweigh the costs. Removal of secrecy would at once reveal how little substance there is in the more grandiose parts of the program. Exaggerated claims of the proponents would be deflated, and exaggerated fears of the opponents would be alleviated. Soviet authorities would be able to assess the program realistically and to see how little it really threatens them. And finally, after the absurdities had been dropped, the more modest and sensible parts of the program would have a better chance to go forward effectively.

The parts of the Star Wars enterprise which are scientifically the most exciting are unlikely to be militarily useful. I am thinking here of the development of new kinds of lasers and of X-ray lasers in particular. These activities ought to be supported on their merits as pure science, without secrecy, in open competition with other scientific projects. The young people who work with furious dedication and skill to explore a new scientific frontier should be encouraged to do so without inventing for themselves a spurious military mission. These young people are excellent scientists but they are mediocre strategists.

In every strategic defense system there are three main components. The first is the tracking and discrimination apparatus, the radars and optical sensors which are supposed to find and identify the targets. The second is the data-handling system, which takes the information from the sensors and feeds it to the computers which launch and steer the interceptors. The third is the interceptor system, the rockets or other more exotic weapons which actually hit and kill targets. The first two jobs, discrimination and data handling, are by far the hardest part of the problem of defense. The third job, sending up an interceptor to kill a target once you know exactly where it is, is comparatively easy. The public discussion of the Star Wars program is mostly concentrated on the easy part of the job, the design and testing of interceptors, and ignores the more difficult parts, the discrimination and information-handling systems.

After we have discarded the technical nonsense and separated the good science from the military nonsense, the remaining solid core of the Star Wars enterprise will be a rather conventional missile-defense research program, emphasizing target discrimination and data processing rather than exotic weaponry. My judgment of what such a program can be expected to do for us agrees with the judgment expressed by Air Force General Jasper Welch a few weeks after President Reagan's announcement. The following paraphrase of General Welch's remarks is a statement of my own position, not of his. He is not responsible for the statement since I am quoting it from memory.

"Look at what it would take to provide a strategic defense of the United States, and look at what we actually have available. The difference between what we need and what we have is about a factor of a thousand. If the Russians threw ten thousand warheads at us and we defended ourselves with the technology we now have available, we might expect to shoot down about ten of them. The Strategic Defense Initiative is supposed to give us the technology to make up the difference between ten and ten thousand. I am not saying that this is impossible, but it doesn't look reasonable. I don't expect any new technology to come along which would really improve things by a factor of a thousand. On the other hand, I can imagine that new technology might improve things by a factor of thirty. And I can imagine that arms control might be successful in reducing the offensive threat by a factor of thirty. If we could get a factor of thirty from strategic defense and another factor of thirty from arms control, we would have a factor of thirty times thirty which is near enough to a thousand. Strategic defense is too big a job for technology to do alone, but it is not too big a job for technology and arms control to do together."

These remarks of General Welch struck me at the time as being the only statement about Star Wars which came from inside the government and still made sense. One might also formulate an argument discussing the prospects of arms-control negotiations in terms similar to those used by General Welch to discuss defense. "Look at what it would take to negotiate ourselves out of the threat of nuclear annihilation," one might say, "and look at what we have actually been able to negotiate. The difference between what we need and what we have accomplished is about a factor of a thousand. We have succeeded only in limiting the numbers of weapons deployed by the United States and the Soviet Union to about ten thousand on each side. To free mankind from the threat of annihilation, we need to negotiate the numbers of weapons down to zero, with some assurance that the number of hidden and uncontrolled weapons does not exceed ten. The arms-control negotiations are supposed ultimately to bring us down all the way from ten thousand to ten. I am not saying that this is impossible, but it doesn't look reasonable. I can imagine bilateral negotiations between the United States and the Soviet Union reducing deployments by a factor of thirty, so that there would be a few hundred warheads left on each side. But it is difficult to imagine negotiation by itself going further than that. When you have a few hundred weapons on each side, the problems are no longer bilateral but multilateral. You have to negotiate with the Chinese and the British and the French as well as with the Russians. And you have to worry more and more about hidden stockpiles and possible violation of agreements. On the other hand, if we could negotiate the offensive forces down by a factor of thirty and at the same time install a reasonably effective strategic defense, I can imagine that the defense might give everybody enough confidence to negotiate an agreement taking offensive weapons all the way down to zero. The transition to a non-nuclear world is too big a job for arms control to do alone, but it is not too big a job for arms control and defensive technology to do together."

I put the argument in this way to emphasize the fact that there is a symmetry between strategic defense and arms control. Strategic defense and arms control are not alternatives. They should be allies. If you want either strategic defense or arms control to get rid of nuclear weapons, you had better have them both together.

My view of the necessary symbiosis of arms control with defense is shared neither by my friends in the professional arms-control community nor by many of the proponents of strategic defense. There is a symmetry here too. The view of the arms-control community has been expressed in a thoughtful booklet, *The Reagan Strategic Defense Initiative: A Technical, Political and Arms Control Assessment,* by Sidney Drell, Philip Farley, and David Holloway of the Stanford Center for International Security and Arms Control. The Stanford experts are bitterly hostile to the Strategic Defense Initiative, partly on technical grounds but primarily on political grounds, because they see strategic defense as incompatible with serious pursuit of arms control. The antagonism between strategic defense and arms control is for them an unquestioned axiom. Their view of the strategic world is dominated by the 1972 Anti-Ballistic Missile Treaty, which is for them the only possible basis for arms control. Anything which threatens the permanence of the ABM Treaty is for them a threat to stability and to peace. I agree with them that the ABM Treaty is worth preserving until such time as the United States and the Soviet Union can agree to replace it with something better. But I have no difficulty in imagining a better treaty, for example, a treaty which allows a build-up of strategic defense while imposing major reductions of offensive weaponry.

On the other side, the enthusiasts for strategic defense are equally uncompromising. If the ABM Treaty has become a sacred cow for the arms-control community, high-tech space weaponry has become a sacred cow for the strategic defense community. The President in his original announcement emphasized the connection of the Strategic Defense Initiative with arms control and said nothing about space. In the technical follies that are now growing out of the initiative, arms-control objectives are forgotten and space weaponry is dominant. Much of the propaganda of strategic defense describes the program as a high-tech arms race in which the chief purpose is to beat the Soviet Union with technology. If that is the chief purpose, then the Stanford experts are right in saying that it is incompatible with serious arms control. The only possible basis of serious arms control is equality. The Soviet Union will not negotiate away its offensive weapons while accepting a technological beating in the area of defense.

The Strategic Defense Initiative now hangs in the balance. Whether it can be a step on the road to a safer world depends on how it is perceived by the Soviet Union. If it could be, as President Reagan's first announcement described it, a slow and long-range program of exploration conducted within limits set by concomitant progress of arms control, there is no reason why the Soviet Union should feel threatened by it. If, on the other hand, it is a mad rush to occupy the high ground of space with dazzling death-rays and other military hardware, then the Soviet Union will never be comfortable with it. The program as it now exists is in reality closer to the President's original concept than to the popular image created by Star Wars propaganda. Unfortunately, the Soviet Union must pay attention to the propaganda as well as to the technical reality.

If you are trying to defend a country against nuclear missile attack, there are two essentially different ways to go about it. The two ways are called by the experts "terminal-phase" and "boost-phase" defense. Terminal-phase defense means that you shoot at missiles as they come in over your own territory. Boost-phase defense means that you reach out over the opponent's territory and try to shoot at his missiles on their way up. There is an enormous difference, technically and politically, between these two kinds of defense. Technically, terminal-phase defense is a straightforward extension of the ABM or anti-missile defense systems which have been developed in the United States and the Soviet Union during the last thirty years. The components of a terminal-defense system are familiar objects such as radars, infra-red telescopes, computers, and interceptor rockets. They are mainly deployed on the ground within the territory of the defended country.

A boost-phase defense is supposed to be built out of exotic and unfamiliar components such as orbiting interceptors, high-powered lasers, orbiting mirrors, particle beams or X-ray beams. A large part of the hardware in a boost-phase defense system must be deployed in space, in orbits which pass constantly over the territory of the opponent. Roughly speaking, a terminal defense is supposed to put a missile-proof shield over one's own territory, while a boost-phase defense is supposed to put a missile-proof ceiling over the opponent's territory.

Politically, a shield is very different from a ceiling. A shield is an assertion of sovereignty over one's own country. A ceiling is an assertion of sovereignty over someone else's country. A shield is a mechanism for keeping intruders out of a country. A ceiling is a mechanism for putting intruders into a country. I can easily imagine the Soviet political system accepting a situation in which both the Soviet Union and its adversaries are protected by more-or-less effective terminal-defense shields. I cannot imagine the Soviet political system accepting a situation in which hostile spacecraft are constantly overhead, ready at a moment's notice to plunge into Soviet territory or to interfere physically with Soviet military facilities in the Soviet heartland. This is the fundamental political difference between terminal defense and boost-phase defense. Terminal defense does not strain the traditional concept of national sovereignty; boost-phase defense is inherently intrusive and must to some extent challenge the sovereignty of the country against which the defense is deployed.

The Star Wars enterprise in the United States includes a variety of projects, some directed toward terminal defense and some toward boost-phase defense. The two types of project have very different characteristics. The terminal-defense projects are real. They are mostly inherited from the old Ballistic Missile Defense program of the United States Army, which has been steadily pushing ahead with terminal defense technology for almost thirty years. The "Star Wars" administration is continuing these projects without any drastic changes. They constitute the solid core of the Star Wars enterprise. The projects directed toward boost-phase defense have an entirely different status. They are new, they are grandiose, they are poorly defined, and as military systems they do not have any real existence. The boost-phase projects are described at military briefings with elegant pictures of nonexistent apparatus. They are a package of vague promises without military substance. But these foggy boost-phase projects have attracted far more public attention than the serious and unspectacular terminal-phase program. Especially in the Soviet Union. The Soviet authorities see pictures in American magazines showing marvelous high-powered lasers sailing over Soviet territory and zapping Soviet missiles. If a laser in orbit over the Soviet Union were in fact capable of destroying Soviet missiles in boost phase, then it would be equally capable of frying the members of the Politburo when they are taking their vacations on the beaches of the Black Sea.

The Soviet authorities are naturally determined not to let anything of this sort happen. They reminded us in 1983, when they shot down the Korean airliner near Sakhalin, that they have a low tolerance for any activity over Soviet territory that might be construed as hostile. It is evident that they will not tolerate the existence over their territory of any apparatus resembling the American magazine pictures. If ever such apparatus should be deployed over Soviet territory, the Soviet authorities have the political will and the technical ability to destroy it. The greatest folly of the Star Wars enterprise is the loud talk about boost-phase projects which nobody knows how to implement, and which, if they were implemented, the Soviet Union would know how to destroy.

The intense opposition of the Soviet Union to the Star Wars program is a natural reaction to our statements about the boost-phase component of the program. Whether or not boost-phase defense is technically feasible, it is politically provocative. Even nonexistent and technically impossible boost-phase projects, when loudly advertised in the public media, are politically provocative. It is just as well that the Soviet authorities have made it clear from the outset that such projects are politically unacceptable. I cannot claim any inside knowledge of Soviet policies or of Soviet thinking. In my interpretation, the integrity of Soviet control over Soviet territory is the decisive factor. Boost-phase defense is unacceptable because it threatens that control. If this is so, then one should not expect the Soviet Union to be unalterably opposed to improvements in terminal defense. The Soviet Union has a terminal defense deployed around Moscow, and is conducting a vigorous program of development of terminal-defense technology. The Soviet Union accepted without protest the American terminal-defense development programs which existed for many years before they were taken over by the "Star Wars" administration. I therefore believe that the terminal-defense component of the Star Wars enterprise would not have been upsetting to Soviet sensibilities if it had not been coupled with wild and exaggerated talk about boost-phase systems. Terminal-defense systems belong in the mainstream of Soviet military traditions. I do not see any reason why the terminal-defense component of Star Wars should not in the long run be politically acceptable to all parties.

That is all I have to say about the political aspect of Star Wars. I now move on to the technical aspect, beginning with some general remarks about the long-term technological balance between attack and defense. The quick German victories at the beginning of World War II created an impression that in modern warfare the attack has an overwhelming advantage over the defense. The same impression was created by the lightning Israeli campaigns of 1956 and 1967. Nevertheless, both in World War II and more recently in the Middle East, the advantage of the attack diminished as time went on. As a result of the development of technology, warfare becomes more and more a battle of information rather than a battle of firepower. As weapons become more accurate, the advantage goes to the side which knows more exactly where the enemy forces are. Accurate weapons without good information are useless. And, as the importance of information increases, we see a gradual shift in the balance of advantage from attack to defense. In a battle of information, the defense fighting from hidden positions within its own territory has an inherent advantage over the attack fighting from exposed and vulnerable vehicles.

A remarkable example of the decline of the advantage of the attack toward the end of World War II is provided by the defense of Finland against the last great Soviet offensive of June and July 1944. I learned the details of this campaign from Stefan Forss, a physicist at the University of Helsinki. The Soviet Marshal Govorov was given thirty divisions of infantry with heavy tanks and artillery, and about one thousand tactical aircraft, to knock Finland out of the war. He was told by Stalin that he had to be in Helsinki in six weeks. After that his divisions would be moved south to join the Soviet armies fighting their way into Germany. There was a desperate battle in Finland, with massive Soviet attacks breaking through three

Finnish lines of defense in succession. The Finns were heavily outnumbered, but had superior information. They had broken the codes used by the Soviet forces for communication by radio. As a result, the Finns usually knew in advance where the Soviet forces were to be concentrated before an attack. The Finnish defense, using carefully placed artillery strikes, could take advantage of this knowledge to delay the attacks and to limit the breakthroughs.

The decisive moment came after six weeks of terrible fighting, when the Finns were barely holding their fourth defense line and the Soviet troops were still 100 miles away from Helsinki. The Finnish code-breakers picked up a personal order from Stalin to Govorov, announcing that the time had run out and that five Soviet divisions were to move south immediately. Govorov replied that he could be in Helsinki in two more weeks. After an hour an angry message from Stalin told Govorov that the war would be decided in Berlin, not in Helsinki. So the five divisions moved south and Finland was saved from sharing the fate of Esthonia, Latvia and Lithuania. Two months later an armistice agreement was signed, surrendering to the Soviet Union the parts of Finland overrun by Soviet forces, and a little bit more, but otherwise leaving Finland intact.

In the forty years which have elapsed since the Finnish campaign ended, the balance of forces in Europe has not essentially changed. The Soviet Union still has a preponderance of tanks and guns and readily available divisions. But any one of the Western European countries, if directly challenged by a Soviet invasion, could perhaps do as well as Finland in defending itself with non-nuclear forces. The development of accurate computer-controlled anti-tank and anti-aircraft weapons has made the job of the defense easier rather than more difficult. As always, the great unknown quantity in such calculations is the morale of the defenders. Would the soldiers of a Western European country be willing to fight as resolutely as the soldiers of Finland fought in 1944? Nobody knows the answer to this question. Fortunately, the Soviet generals also do not know the answer.

The main conclusion which I draw from the example of Finland and from other examples of countries which have successfully defended their independence is that the defenders do better without nuclear weapons. It is difficult to imagine any way in which a country could use nuclear weapons effectively in a defensive campaign against Soviet forces. The use of nuclear weapons against the Soviet Union would be a suicidal action for the country that used them. And the deployment of nuclear weapons in peacetime gives the defending troops the impression that their function is unreal. An intelligent young man who is called upon to defend his country as a soldier in a nuclear-armed army cannot be expected to take his job seriously. He must know that in case of any serious fighting the army to which he belongs, and probably the country to which he belongs, will cease to exist. When soldiers are trained for a job that they believe to be unreal, we cannot expect their morale to be high. If the NATO Alliance has doubts about the morale of its troops, one of the necessary steps toward improving their morale would be to reduce their dependence on nuclear weapons, to make it possible for them to defend their country without destroying it.

But I am digressing from my theme. The theme is the gradual shift in military advantage from attack to defense as warfare becomes less a matter of crude firepower and more a matter of accurate information. The defense of Finland in 1944 was one illustration of this theme. The possible non-nuclear defense of Western Europe is another. A third instance of the same theme is the Star Wars problem, the problem of the strategic defense of the United States and the Soviet Union against nuclear attack. The orthodox view among the experts is that in the arena of strategic nuclear weaponry, the supremacy of attack over defense is permanent and unchallengeable. Nevertheless, I am challenging it. I am saying that even in this most difficult kind of defense, the defense of a fragile human society against weapons of mass annihilation, the defense may in the long run prevail if it takes full advantage of the information available to it. All that is needed to destroy a nuclear warhead is to put half a pound of iron in the right place at the right time. The half pound of iron can be carried by a small and light interceptor rocket. To kill a warhead does not require any high-technology laser beams or electromagnetic guns. All it requires is information. The defense must know accurately where the warhead is, and must be able to convert this information rapidly into instructions for the accurate guidance of the interceptor. And the apparatus of defense must itself be dispersed and hidden so that the attack is unable to seek out and destroy it.

It is possible to imagine that, as the technology of sensors and microcomputers and data processors improves, a terminal-defense system making full use of all the available information may finally gain an advantage in cost effectiveness over the attack. There are many reasons why improvements in information technology favor the defense, provided that the defense is terminal. The defensive battle is to be fought over the defended territory in full view of the defenders, whereas the attack will be fighting blind from bases thousands of miles away. The defensive rockets can be quick and agile, whereas the attacking warheads are heavy and relatively unmaneuverable. The defensive system can be effectively concealed in the ground, whereas the attacking forces are necessarily exposed in the transparency of space and are even more visible as they reenter the atmosphere. For all these reasons, the future of terminal defense looks technically promising. Non-nuclear defense against nuclear attack will never be perfect or reliable, but it may become technically good enough to make multilateral nuclear disarmament possible. That is the announced goal of the Star Wars initiative. To reach the goal, the defense must take maximum advantage of its easier access to information. The advantages of the defense lie in its better view of the battlefield, in its shorter lines of communication, and in the protection afforded by ground cover.

All these technical advantages of defense belong to terminal defense only and are lost or reversed as soon as we move to a boost-phase defense. So far as access to information is concerned, a boost-phase defense reverses the roles of attack and defense. In a boost-phase system, it is the defense which is exposed and vulnerable, the defense which has a poorer view of the battlefield and longer lines of communication. These disadvantages of boost-phase defense appear to be inescapable, and independent of the details of the hardware. If there is any boost-phase defense technology that avoids these vulnerabilities, it can only be some modernized version of the old BAMBI system proposed in the 1950s, with an enormous number of very small autonomous interceptors dispersed in orbits around the Earth. Even this "swarm of bees" approach to boost-phase defense can probably be nullified by ground-based countermeasures. Any boost-phase defense using larger and more sophisticated units in orbit will be even more vulnerable. It is superfluous to discuss whether a particular high-powered laser system might or might not be effective in disabling a Soviet booster over Siberia. Whether or not the laser were effective, the optical apparatus which aimed the laser beam into Siberia would necessarily be far more vulnerable to attack than the booster. A rather simple terminal defense on the ground in Siberia could put out of action the apparatus of boost-phase defense overhead.

The conclusion of this discussion of strategic defense is clear-cut. The political part of the discussion led to the same conclusion as the technical part. The boost-phase defense systems which, if they could be built, would be politically intrusive and unacceptable to the Soviet Union, are also technically unpromising. They are inherently vulnerable to technical countermeasures and are likely to become less and less viable as information technology improves. The terminal-defense systems which might be politically congenial to the Soviet Union are also the systems which have technical promise. They are likely in the long run to become technically attractive as the technology of sensors and computers moves ahead.

So long as the Star Wars program emphasizes boost-phase defense, it is a dead end. It can lead to no militarily useful output, and it ensures Soviet opposition to any arms-control agreement which allows boost-phase defense to move toward deployment. On the other hand, a Star Wars program reduced to an exploration of terminal-defense technology might lead to something technically sensible and politically negotiable. It will not be easy to salvage anything useful from the program. The web of illusions and fears that has grown up around the idea of boost-phase defense will not be easily swept away. But I still hope that we and the Soviet authorities may have the wisdom to make a distinction between boost-phase and terminal-phase defense. Exploration and improvement of terminal-phase defense may in the end be helpful to us all in creating a basis for a stable balance of power while nuclear weapons are being withdrawn or eliminated. One may hope that this possibility may be kept open when the fantasies of boost-phase defense are abandoned. That is my verdict on Star Wars. The terminal-phase component of Star Wars should be given back to the United States Army and continued as a military program, like the Soviet ballistic missile defense program, in a sober style without exaggerated claims. The boost-phase components of Star Wars should be abandoned as military programs, and continued as non-military projects to the extent that their scientific interest warrants.

It should be possible to conduct the Star Wars program in a style that Soviet leaders will find acceptable. The aim of the program should not be to outrun the Soviet Union but to chart a path which the Soviet Union can prudently follow. Fortunately, the boost-phase part of the program, which is the part most objectionable to the Soviet Union and the part most unlikely to be militarily useful, is also the most expensive. There is a good chance that the United States Congress will trim the program's budget sufficiently to bring it down to Earth. If the program is conducted modestly, with a minimum of extravagant claims and a maximum of sensitivity to Soviet anxieties, it may in the end fulfill the purpose which President Reagan expressed in his original announcement.

President Reagan has suggested that if the Star Wars program were successful, we should offer to share the resulting defensive technology with the Soviet Union. This suggestion was widely regarded as naive and unrealistic. It is indeed difficult to imagine either American or Soviet political systems accepting gracefully a collaboration between Livermore and Sary Shagan. But the essential purpose of the President's suggestion of sharing would be achieved if our program were declassified. Declassification yields the benefits of sharing without the political difficulties. And declassification has the important advantage of keeping the Soviet Union informed when our program fails, not only when it succeeds. On both sides, secrecy conceals failure more often than it conceals success.

The purpose of strategic defense is not to save our skins in case we get into a nuclear war. That is why the arguments which prove that defense is imperfect and uncertain do not prove that it is useless or undesirable. The purpose of defense is to create a state of mind. The purpose is to persuade political and military leaders all over the world that nuclear weapons are not a good buy. The purpose is to make nuclear weapons unnecessary and unattractive. President Reagan said that he would like to make nuclear weapons "impotent and obsolete." That was an unfortunate choice of words. It promises too much and demands too much of the defense. If I had been writing the President's speech, I would have said "unnecessary and unattractive." The defense has done its job if it makes nuclear weapons unnecessary and unattractive, so that it will make sense to think of getting rid of them altogether.

13

THE EXAMPLE OF AUSTRIA

Strategic defense is not the most important issue facing the world today. Perhaps the chief damage which President Reagan's "Star Wars" initiative has done is to distract our attention from other more important matters. The more important matters are political rather than technical. They concern people on the ground rather than weapons in the sky. The most important question, as Lenin used to say, is "Kto kogo?" ("Who whom?"), who gives the orders and who obeys them. The political antagonisms which drive arms races are more important than the details of the weapons.

One of the more important matters on my agenda is the political future of Germany. One of the possible futures of Germany is Austrianization. "Austrianization" is an ugly word but it stands for a hopeful concept. The essence of the concept is the idea that the Austrian State Treaty of May 1955 was the greatest achievement of international arms control in the years since World War II, far more important than the ABM Treaty and the other treaties which deal with weapons rather than with people. Until 1955, Austria had been divided like Germany into zones of occupation, with Soviet troops in the Eastern zone and American, British and French troops in the Western zones, with a divided regime in Vienna similar to the divided regime in Berlin. The 1955 treaty abolished all this, got the Soviet and Western troops out of the country, and established Austria as a free and independent state. Anybody who goes to Austria today can see the consequences of the treaty. Austria is now an astonishingly prosperous, politically stable and Western-oriented country.

No matter how strongly the concerned public in other countries may disapprove of the character of Kurt Waldheim, Austria elected him president and thereby demonstrated a sturdy contempt for foreign political pressures. Austria occupies a strategic territory in the heart of Europe, sticking out far to the east of the Soviet satellite capitals of Prague and Berlin. And in the thirty years that have passed since the State Treaty was signed, there has been no sign of Soviet dissatisfaction with it. The Soviet government has never threatened to invade Austrian territory or to interfere seriously in Austrian internal politics. This happy state of affairs raises several questions. What was the price paid to achieve it? Why did the Soviet government accept it? Would a similar price be sufficient to buy us similarly happy results elsewhere?

Although the United States and Britain and France played an important part in the negotiations, the chief credit for the existence of the Austrian State Treaty belongs to the Austrians themselves. The essential conditions of the treaty were decided in bilateral negotiations between the Austrian chancellor Julius Raab and the authorities in Moscow. The Austrians, unlike the Germans, had maintained an internal political unity under the four-power occupation, so that Raab could speak with authority for the whole country. And Raab knew what he wanted. He wanted the Russians out and was prepared to pay any reasonable price to get them out. The price, as it turned out, was not excessively high. First, a substantial sum in cash was demanded as payment for the handing over of German assets which the Soviet government considered to be Soviet property. Second, the Austrian state had to write into its constitution a pledge of perpetual neutrality, which was duly inserted a few months later. This means that Austria is forbidden to join NATO and is required to defend its independence against all threats, in particular against any attempt by Germany to repeat the Anschluss of 1938. In addition to the cash payment and the pledge of neutrality, there were various minor conditions. The Austrian military forces are forbidden to employ soldiers who held a rank of colonel or higher in Hitler's armies. Employment of foreign mercenaries is forbidden. No restrictions are imposed on numbers of troops. All things considered, it was a remarkably lenient settlement for a country which had welcomed Hitler's rule in 1938 and had been a full partner in the invasion of the Soviet Union in 1941.

We shall never know precisely why the Soviet authorities agreed to the settlement. From the Soviet point of view, the immediate advantage of the treaty was that it pushed back the border of the NATO Alliance by 100 miles, from Austria to Bavaria. Nevertheless, it remains a paradox that the Soviet Union was willing to tolerate in Austria a politically free society with strong cultural and economic ties to the West, while suppressing rigorously any movement toward similarly free societies in Czechoslovakia and Hungary. Why should a neutral Austria be perceived as less threatening to Soviet power than a neutral Czechoslovakia? The key to Soviet actions all over Europe is probably to be found in Germany. The generous treatment of Austria was probably intended as a signal to the population of West Germany, demonstrating the possibility of a happy and prosperous future outside the NATO Alliance. Chancellor Konrad Adenauer stood firm against Soviet blandishments and led West Germany into full membership of NATO in the same month in which the Austrian State Treaty was signed. Nevertheless, the Austrian example still stands as a reminder to the younger generation of Germans of the possible future which their elders rejected.

It happens that I am married into a German family, and I have three nephews who are West German citizens. Through the nephews I have an impression of the world as seen through the eyes of young Germans. They are acutely aware of the presence of Soviet troops in occupation of the eastern part of their country. They are not in any sense pro-Soviet or pro-Communist. But they are deeply dissatisfied with the policies of the NATO Alliance. They are opposed to the stationing of nuclear weapons in their country, and they have no confidence in the wisdom of the American leadership of the alliance. They look at the situation of their neighbors across the border in Austria, and they say, "Why not us?" I expect this question to be heard more and more insistently in Germany as the younger generation takes over the responsibilities of power. Austria, it seems, is doing very well without nuclear weapons, without American allies, and without a Soviet army of occupation.

The idea of neutrality as an alternative policy for Germany is not a new one. In the past the opponents of neutrality have introduced the word "Finlandization," which is supposed to imply that a neutral Germany would find itself in a relation of political subservience to the Soviet Union. The word "Finlandization," when used in this sense, is a slanderous misrepresentation of the actual state of affairs in Finland. Finland has stoutly and successfully defended its independence against Soviet invasion, and it remains, like Austria, a free and Western-oriented society. Finland is not a Soviet satellite state. Nevertheless, Finland's geographical situation, with its long land frontier directly adjoining major centers of Soviet naval and industrial power, and Finland's history as the only fully independent state existing in the territory of the old Russian Empire, make it inevitable that Soviet anxieties remain an important factor in Finnish political life. The first rule of Finnish foreign policy is to do nothing that would endanger the maintenance of peaceful coexistence with the Soviet Union. Only in this very limited sense is Finland subservient to Soviet power. The situation of Austria is, both for geographical and for historical reasons, much easier. Austria was never Russian territory, and Austrian troops are not within shooting distance of Leningrad and Murmansk. In these respects the situation of a neutral Germany would be more analogous to the situation of Austria than to that of Finland. I therefore propose that we speak not of Finlandization but of the Austrianization of Europe.

The Austrianization of Germany would transform the political equilibrium of Europe in ways that we may or may not consider desirable. Whether we like it or not, we ought to be thinking about it. Whether we like it or not, the United States may one day wake up to find that the German people have voted for Austrianization and that we have no power to prevent it. It would be to our advantage to discuss the conditions of Austrianization with the Soviet Union before the German governments take matters into their own hands. It is not too soon to begin thinking how we might use the possibility of Austrianization to achieve two of the chief aims of NATO policy, the withdrawal of Soviet troops from Germany and the removal of Soviet nuclear weapons from the European theater. Whether the Soviet Union would accept these conditions, we cannot say. My German nephews would accept them enthusiastically. We may hope that a new generation of leaders in Germany and in the United States may agree that the withdrawal of American troops and nuclear weapons from Germany is a reasonable price to pay for corresponding withdrawals on the Soviet side. A German State Treaty along the same lines as the Austrian treaty would provide an opportunity for both sides to withdraw.

There is an important difference between Austria and Germany. Germany has two governments, and the Austrianization of Germany does not require unification of the governments. For many reasons, both internal and external to Germany, it would be wise to leave the two governments in existence when the treaty comes into force. The possible unification of the governments and of the city of Berlin might or might not be left as a subject for later negotiation. It may well be that the German population, like the populations of Norway and Sweden after the separation of their governments in 1905, will find that they can live with two governments more comfortably than with one. The major hardship which the present division of Germany imposes on the population is not the existence of two governments but the barriers to communication and travel across their border.

What then does the Austrianization of Germany mean? Here are two possible versions of an Austrianization treaty. Version A has five clauses, Version B has six. Version A is as follows. First, withdrawal of foreign troops from German territory, East and West. Second, withdrawal of the two German governments from their respective alliances. Third, constitutional pledges of permanent neutrality of both governments. Fourth, prohibition of nuclear weapons and nuclear delivery systems from German territory. Fifth, the two German governments are pledged to defend their sovereign independence and are forbidden to unify. Version B has clauses one to four the same as Version A, but instead of clause five it has: Fifth, prohibition of the use of force to unify Germany. Sixth, in case the two governments peacefully agree to unification, all obligations under the treaty devolve upon the successor government. The essential difference between the two versions is that Version A establishes the German Democratic Republic as a permanently legitimate and sovereign state, just as the Austrian State Treaty did for Austria. Version B leaves the status of the German Democratic Republic ambiguous. When I discussed these proposals with some Soviet officials who were visiting Princeton recently, they expressed a strong preference for Version A. I have also found a preference for Version A among my American and West European friends. Richard Ullman, a professor of political science at Princeton, expressed vividly their misgivings about Version B: "You don't want to leave Germany as a loose cannon on the deck of a rolling ship." On the other hand, Version B would be preferred by the majority of West German voters.

The two versions A and B are not the only possible alternatives. It is easy to invent further variations on the same theme. One might compromise the unification issue by permitting the two German governments to unify peacefully after fifty or a hundred years have gone by. The treaty might or might not regulate the size and equipment of armies belonging to the two governments. It would be wise to make the treaty as simple and as short as possible. The main reason for separating the issue of neutralization from the issue of reunification is that reunification is bound to be complicated while neutralization is comparatively simple.

The details of an Austrianization treaty will depend on circumstances which cannot be foreseen. I want now to stand back from the details and look at the thing in a larger perspective. The treaty would be a revolutionary development in European politics, because it would be a major move away from military confrontation and toward a new international order. Almost incidentally, it would sweep aside thousands of the most dangerous short-range nuclear weapons. The short-range weapons would have no place to go after the NATO and Warsaw Pact armies were withdrawn from nose-to-nose contact. But the larger significance of the treaty would be the building of a new international order to replace the direct military domination of Europe by the two great powers.

Opponents of Austrianization will say that it makes the survival of freedom in Germany dependent on Soviet goodwill. This is true, just as it is true that the Austrian State Treaty made the survival of freedom in Austria dependent on Soviet goodwill. The Austrian example shows that freedom under these circumstances can be quite robust. The word "goodwill" does not describe completely the Soviet attitude toward Austria. The Soviet acceptance of Austrian independence is based not so much on goodwill as on a commitment to law and order in international relationships. Soviet acceptance of German independence after neutralization would likewise depend on Soviet commitment to a new international order in Europe. I wrote a long chapter in my book *Weapons and Hope* describing the historical task of diplomacy as the building of an international order on the basis of a realistic balance of power. The balance of power is essential if the international order is to be stable. But the balance of power in Europe does not require deployment of American and Soviet troops in Germany. After German neutralization, the balance of power would still exist.

The Soviet Union would have as strong an incentive to behave correctly toward Germany as toward Austria. The balance of power resides in the roughly equal massiveness and mobility of American and Soviet resources, with French and British forces approximately compensating for the geographical remoteness of America. The balance of power in Europe is fortunately stable and likely to be permanent. It would provide a solid basis for a new European order with two neutral German states in the center. The Soviet Union would still know that the price of any invasion of Austrian or of German territory would be an acute danger of major war with all its incalculable consequences.

If we imagine that the Austrianization of Germany could be successfully accomplished, we may allow imagination to spread further and suppose the process of Austrianization extending gradually over the rest of Europe. Sweden, Finland, Switzerland and Yugoslavia are already enjoying the benefits of neutrality. It could happen that Austrianization will spread to Norway, Denmark and Italy on the NATO side and to Hungary, Czechoslovakia and Poland on the Warsaw Pact side of Europe. All this will take a long time and much patience. But it is a hopeful road to follow, once we can get started. The essential condition that would make it possible for the Soviet Union to allow Austrianization of Eastern Europe is the removal of the threat to Soviet security arising from nuclear-armed and American-allied Germany. We cannot know whether the Soviet Union will ever let us reach the goal of a completely Austrianized and denuclearized Europe. It will do us no harm to hope and to try.

The foregoing paragraphs were written in 1985. I began three years ago to give this sermon about Austrianization when I was asked to talk to local church meetings in America. The future of Germany was a good subject for discussion among groups of people concerned about problems of war and peace. I put forward the idea of Austrianization because my audiences needed a change from endless technical arguments about details of the Star Wars program. I did not expect the idea of Austrianization to be welcomed by the experts. The idea came to me from Harald Jung, the youngest of my three German nephews, who came to stay with us in Princeton three years ago. Harald was then only fourteen, and perhaps wiser than most of his elders. He was able to look at the problems of the world with fresh eyes. I thought his ideas made sense and so I adopted them as my own.

To my astonishment, I found that some of the political experts in Princeton were taking these ideas seriously. A few weeks later I was invited to give my Austrianization talk to an audience in New York which included the Austrian consul-general and various other dignitaries. The dignitaries mostly disagreed with what I said but at least they listened. The consul-general gave me a copy of Gerald Stourzh's history of the Austrian State Treaty. This is a scholarly account of the tangled negotiations which continued intermittently for eight years until they finally resulted in the 1955 treaty. After I had read the Stourzh book I could fairly claim to be an expert. An expert in any subject is somebody who has already made all possible mistakes. From the Stourzh book I learned that a number of statements in my talk were factually wrong.

The most surprising fact which I learned from a careful reading of the text of the Austrian State Treaty is that the neutrality of Austria is nowhere mentioned in it. The treaty was signed on May 15, 1955, and came into force after ratification by the five signatory countries on July 27. The last contingent of occupation troops left Austrian soil on October 25. The last contingent was, as it happened, British, and the Austrians in Klagenfurt gave a big farewell party for the departing colonel in charge. One day later, on October 26, the neutrality of Austria was established by a constitutional law passed by the Austrian parliament. The Austrians had insisted from the beginning that things must be done in this order, first the treaty, then the withdrawal of occupation forces, then the neutrality law, and they had their way. A neutrality law, they said, could only stand as a perpetual and solemn commitment of the Austrian people if it was a free act of a sovereign Austrian state. And so it was done. The State Treaty, signed while Austria was still an occupied country, contains no such commitment.

The neutrality law is an admirably short and simple document. It consists of two articles. Here is the complete text:

> *Article 1.* For the purpose of the lasting maintenance of her independence externally, and for the purpose of the inviolability of her territory, Austria declares of her own free will her perpetual neutrality. Austria will maintain and defend this with all means at her disposal. For the security of this purpose in all future times Austria will not join any military alliances and will not permit the establishment of any foreign military bases on her territory.
>
> *Article 2.* The Federal Government is charged with the execution of this Federal Constitutional Law.

It is this law which I would recommend as a model of clarity and brevity for the two German governments to emulate. I was wrong when I said that the Austrian State Treaty was short and simple. The treaty contains thirty-eight articles, with a long preamble, two annexes, and an appendix. Most of it is concerned with detailed regulations for the disposal of German military and economic assets left behind in Austria at the end of the war. These complicated arrangements, the result of years of strenuous diplomatic argument, are now of little interest even to historians and are irrelevant to the future of Germany. I was right in saying that the treaty imposes no quantitative limits on Austrian troops and weapons. But it does impose qualitative limits. Article 13 sets forth a list of ten categories of weapons which Austria is forbidden to produce or to possess, beginning with nuclear weapons and ending with chemical and biological weapons. Included in the list are guided missiles, torpedoes, submarines, mines, and artillery with range greater than 30 kilometers. This list might or might not be considered appropriate for the regulation of weapons in the two German states. A future German State Treaty should contain a clause analogous to Article 13 of the Austrian Treaty.

The most illuminating part of the Stourzh history is the description of the bilateral negotiation between Austria and the Soviet Union in March and April of 1955. This negotiation settled the main issues. Here for the first time the Austrians offered and the Soviet Union accepted a constitutional neutrality law as a foundation of Austrian independence. The negotiation ended with a formal memorandum, not a treaty, which set forth separately the intentions of the two sides. Both in the Austrian and in the Soviet part of the memorandum it is explicitly stated that Austrian neutrality should be legalized in the same fashion as Swiss neutrality. Swiss neutrality was officially established in international law by the action of the European powers assembled at Vienna on November 20, 1815. Switzerland had written a law of perpetual neutrality into the Swiss constitution, and the European powers at Vienna jointly guaranteed the inviolability of Swiss territory. Austria and the Soviet Union in April 1955 agreed to follow the Swiss pattern. The negotiation in Moscow was to a remarkable extent a dialogue between two men, Raab and Molotov. According to Stephen Verosta, one of the Austrians who was present, Molotov was in command of every detail of the subject matter and hardly ever needed a word of advice from his advisers. Nobody knows how it happened that this man, who for many years had been the embodiment of the unyielding and stone-faced Soviet style of diplomacy, could suddenly transform himself into an agreeable host and an effective peacemaker.

In September 1955, shortly after the signing of the Austrian Treaty, the Soviet Union concluded a similarly reasonable treaty with Finland. Until 1955, the Soviet Union had maintained an occupation force in two areas of Finnish territory, one at Hangö at the entrance to the Gulf of Finland, the other at Porkkala a few miles from Helsinki. The 1955 treaty withdrew the occupation forces and gave these areas back to

Finland. In return, Finland had only to renew the already existing non-aggression pact and commercial agreement with the Soviet Union. When I visited Finland in 1956, I was taken for a drive through the abandoned Soviet base at Porkkala. It was a striking contrast, to go directly from the elegant beauty of Helsinki to the derelict squalor of Porkkala. And still, I was full of admiration for the wisdom of the Soviet authorities who made the decision to withdraw unilaterally from Porkkala. Unilateral withdrawal is not easy for a great power. The United States is still occupying the naval base of Guantanamo in Cuba. The situation of Guantanamo now is in many ways similar to the situation of Porkkala before 1955. Guantanamo is an annoyance to Cuba and an unnecessary expense to the United States. But the United States has never had the wisdom to withdraw.

The behavior of the Soviet Union is always unpredictable. There was a time in 1955 when the Soviet Union took two actions, the withdrawal from Austria and the withdrawal from Finland, which helped greatly to establish conditions of international order and stability in Central and Northern Europe. At other times the policies of the Soviet Union have been less accommodating. The actions of 1955 came at a time when the internal political equilibrium of the Soviet Union was shifting rapidly after the death of Stalin. It is possible that there will again be moments in the future when the Soviet Union is ready to make generous moves toward peace and disengagement. We in the West must be ready to respond fast when such moments occur. That is the lesson of the Austrian State Treaty. Patience was rewarded. The Austrians and the Western allies negotiated patiently for eight years to obtain an acceptable treaty, and then suddenly Molotov smiled and the treaty was signed within two months. This could happen again, if we are ready to move fast and do not miss the opportunity when it comes. It could happen in the future with negotiations for a political settlement in Germany. Or it could happen with negotiations for a general abolition of nuclear weapons. In all these areas, we cannot compel the Soviet Union to move and we cannot predict when the Soviet Union will be ready to move. All we can do, all we must do, is to be ready ourselves with wise and generous proposals whenever the happy moment comes.

14

CAMELS AND SWORDS

I come back again to that distant but inescapable goal, the abolition of nuclear weapons. Most people do not believe that abolition is possible. Many people, especially people who are professionally involved with diplomacy and arms control, do not believe that abolition would be desirable even if it were possible. Recently I was at a meeting of scientists in California. We were mostly physics professors who consider ourselves to be liberal and enlightened. The speaker said: "Let's take a vote. I want you to imagine that you wake up tomorrow morning and hear the news that all the nuclear weapons in all countries have magically disappeared. And imagine that you have certain proof that the weapons are gone. How many of you would feel safer and how many of you would feel less safe?" Without hesitation I held up my hand to vote for feeling safer. Sadly I have to report that the great majority of my friends and colleagues voted for feeling less safe. They honestly believe that nuclear weapons make the world safer. The majority of political experts in the governments of Western Europe and the United States seem to be of the same opinion. They regard nuclear weapons as instruments of international stability and security. Abolition of nuclear weapons would change unpredictably the rules of the game they have learned to play.

Against this consensus of the experts who have learned to love the bomb, a few powerful voices of protest have been raised. Here is a passage from the concluding summary of the Pastoral Letter on War and Peace approved by the Catholic Bishops of America:

> The nuclear age is an age of moral as well as physical danger. We are the first generation since Genesis with the power to virtually destroy God's creation. We cannot remain silent in the face of such danger. Why do we address these issues? We are simply trying to live up to the call of Jesus to be peacemakers in our own time and situation.
>
> What are we saying? Fundamentally, we are saying that the decisions about nuclear weapons are among the most pressing moral questions of our age. While these decisions have obvious military and political aspects, they involve fundamental moral choices. In simple terms, we are saying that good ends, defending one's country, protecting freedom, etc., cannot justify immoral means, the use of weapons which kill indiscriminately and threaten whole societies.

The Pastoral Letter did not have any visible effect on the policies of the American government. Nevertheless, it was an important landmark in the history of nuclear weapons. It was the first time that a deeply conservative hierarchy, traditionally identified with patriotic sentiment and military strength, took a firm position in opposition to the declared nuclear strategy of the United States. Concerning the American strategy of nuclear deterrence, the bishops supported the judgment pronounced by Pope John Paul II himself: "The logic of nuclear deterrence cannot be considered a final goal or an appropriate and secure means for safeguarding international peace."

The abolition of nuclear weapons, if it ever happens, will not happen suddenly or by magic. It is likely to happen gradually, by a series of small decisions rather than by a grand design. If we look back into the past, we can find many occasions on which nuclear weapons have actually been abolished. We used to have nuclear anti-aircraft missiles, Nike Hercules and Bomarc, scattered widely around the United States. They are not there now. We used to have nuclear intermediate-range missiles, Jupiter and Thor, deployed in Italy and Turkey. They are not there now. We used to have nuclear missiles called Subroc widely deployed on American submarines. They are not there now. Furthermore, these nuclear weapons were not merely withdrawn to be replaced by other nuclear weapons. They were in effect abolished. And they were abolished, not for grand ideological reasons but for simple practical reasons. They were abolished because they were more trouble than they were worth. They were abolished because the missions they were supposed to perform didn't make much sense. They were abolished to save money for other things. And that is probably the way it will be in the future. We shall abolish nuclear weapons, not by a sudden outburst of peace and goodwill but by a slow process of erosion. The weapons will be abolished as the missions for which they were designed come to seem unnecessary or absurd. This will take a long time. It will not be easy. But then the Catholic bishops never said that it would be easy.

Nuclear weapons are deeply entrenched, not only in the concrete silos of Montana and Siberia but also in the structure of international relations and in our ways of thinking. It is difficult to imagine the process of transition which could take us from the world of today to a world in which nuclear weapons were no longer important. But the fact that a historic transition is unimaginable before it happens does not imply that it will never happen. History is full of examples of transitions which upset deeply entrenched institutions and deeply held beliefs.

It has happened not infrequently that a dominant technology, accepted by contemporary observers as permanent, disappeared from the scene with astonishing swiftness.

Sometimes a technology disappears because it is replaced by something more powerful: sailing ships succumb to steam and canals succumb to railroads. Sometimes a technology disappears because it is replaced by something less powerful; there may be social or political forces which cause a less powerful technology to prevail. I call a transition from less to more powerful technology a transition of the first kind, and a transition from more to less powerful technology a transition of the second kind. Transitions of the first kind are familiar to us. Transitions of the second kind are rarer and less well known. Since the transition to a world without nuclear weapons must be a transition of the second kind, it is important for us to study transitions of the second kind with special care, in order to understand how and why they have happened.

A striking example of a transition of the second kind has been described by Richard Bulliet in his book *The Camel and the Wheel*. Bulliet is a historian of early Arab civilization. He demonstrates with ample documentation that in Roman times the entire Arab world, extending roughly from Tunis to Afghanistan, based its economic life on the same infrastructure as the Roman Empire, namely, on the technology of wheeled vehicles and paved roads. South and east of the Mediterranean as well as north, the basic unit of freight transportation was the oxcart. About A.D. 500, a few hundred years before the rise of Islam, a drastic change occurred. Throughout the Arab territories, caravans of camels took over the freight business, roads fell into disrepair and wheeled vehicles disappeared. For more than a thousand years, until Europeans moved in with steel rails and locomotives, the camel reigned supreme.

Bulliet was able to identify with some precision one of the crucial events which led to the decline of the oxcart and the victory of the camel. He found a fifth-century tax code from the city of Palmyra, an important trading center in northern Syria. Palmyra levied taxes on all shipments passing through the city. The tax rates were set so that an oxcart paid as much as four camels. But Bulliet could calculate independently the payload of a camel and of an oxcart. It turned out that a camel could carry on the average about 600 pounds and an oxcart could carry about 1,200. An oxcart could carry as much as two camels but paid as much as four camels.

The tax rates were set so as to discriminate against oxcarts. Probably the officials who wrote the tax code had family connections with the camel drivers. In any case, a discriminatory tax code would be effective in tipping the balance of trade against the oxcart. And the balance between the two competing systems of transport was inherently unstable. The commercial efficiency of the oxcart depended on an infrastructure of roads which had to be kept in good repair. As soon as the oxcart business started to decline, the roads would have started to deteriorate. As soon as the roads deteriorated, the decline of the oxcart would have become rapid and irreversible. Within one or two generations there would no longer be any skilled craftsmen who knew how to build and repair oxcarts. Not only the oxcart, but even the memory of its existence, disappeared from the Arab world. The word for a wheeled vehicle vanished from the Arab language for a thousand years.

Another transition of the second kind occurred in Japan in the seventeenth century. It is described in a book, *Giving Up the Gun,* by Noel Perrin. In the sixteenth century, after the first European ships visited Japan, Japanese swordsmiths quickly learned to make guns. They manufactured guns of superior quality. Large numbers of Japanese guns were exported, and still larger numbers were used by Japanese armies abroad and at home. Perrin illustrates his book with old Japanese drawings of seventeenth-century warriors carrying and firing guns. For half a century the corps of Samurai was heavily addicted to guns. Gun battles were fought with great loss of life. The losses were so heavy that the leading Samurai became convinced that guns were ruining their honorable profession. They decided to go back to fighting with swords. For two and a half centuries the sword was reinstated as the basis of military power. The supremacy of the sword was maintained until 1879, when the Samurai were defeated by a new-style army intent upon modernization and armed with European weapons.

Perrin was able to trace in detail the process by which guns were discouraged in Japan, just as Bulliet did for the oxcarts in Palmyra. In both cases, the mechanisms of discouragement were economic rather than legal. Guns in Japan were never formally forbidden. The government merely encouraged their disappearance by making their manufacture a state monopoly and setting prices unreasonably high. Guns continued to be produced in small numbers but were too expensive to be used for any but ceremonial purposes.

The main reason why the Samurai disliked guns was that a gun made a peasant the equal of a Samurai. When guns were in vogue in the seventeenth century, battles were fought by armies of peasants armed with guns. The lifelong training and dedication of a Samurai to his art then counted for almost nothing. If peasants could win battles, what need would a feudal lord have for Samurai? The Samurai despised the peasants, just as the camel drivers of Arabia despised the drivers of oxcarts.

These two transitions of the second kind have much to teach us. In both cases, the driving force of the transition was the political power of a skilled professional elite, the camel drivers in Africa and the Samurai in Japan. In both cases, the ideology of the transition was conservative, aiming to perpetuate an old social order and a traditional way of life. In neither case was the transition permanent, but it lasted more than a thousand years in Africa, more than two hundred in Japan. If we could successfully abolish nuclear weapons and make the abolition stick for two hundred years, that would at least give our species time to adapt itself to new technologies and to tackle some of the other urgent problems of survival.

A practical program leading toward the abolition of nuclear weapons must be based upon three principles. First, it is not possible to replace something with nothing. To replace a wheel we need a camel. Our camel must be a robust and versatile technology of non-nuclear weapons capable of defending our interests and our allies without destroying us. Second, the only political force strong enough to abolish nuclear weapons is the military establishment itself. The move away from weapons of mass murder must be presented to the public, not as a response to fear but as a response to the demands of military honor and self-respect. Third, the military establishment of the Soviet Union must be allowed an equal share with our own in the shaping of the program. The stabilization of a non-nuclear world is a conservative objective, and the hope of achieving it rests mainly on the profoundly conservative traditions of the military professionals in all countries. To succeed, we need to have on our side not only the camel drivers but also the Samurai.

If we are trying to deal with the problem of nuclear weapons, there are important lessons to be learned from Japan. Not only from seventeenth-century Japan but from modern Japan too. In 1985 I went on a lecture tour to Japan. That was my first visit and so I included in the tour a pilgrimage to Hiroshima. When you go as an official guest to Hiroshima, it is like going on a pilgrimage. You arrive in a solemn frame of mind, remembering the great slaughter and the many books that have been written about it, carrying on your shoulders a small share of responsibility for the continued existence of nuclear weapons, carrying in your heart a small feeling of guilt. Coming to Hiroshima, we feel guilt not so much for the slaughter of 1945 as for our persistence in the same habits of thought and action which made that slaughter inevitable and now may lead us, if we are unlucky, to slaughter on an even grander scale.

The official welcome which the city of Hiroshima extends to visiting pilgrims is well designed to encourage a mood of solemn reflection. My wife and I took plenty of time to examine the carefully preserved relics of the holocaust, the little shirts and shoes and exercise books and lunchboxes of the children who died that August morning forty years ago. We drank tea with an old lady who had been teaching English to some of those children. Half of her class survived. She herself was one of the lucky ones, having been dug out of the rubble before the fire-storm had time to sweep over her. She is now retired as a teacher and is devoting herself full time to the cause of peace. She is preaching, to all who will listen, the necessity of a total abolition of nuclear weapons. I already believed, before I went to Hiroshima, that total abolition is a good idea. The days that we spent in Hiroshima made the goal of total abolition seem more practical and less remote.

We had meetings with several groups of Japanese students and professors, and with a group of members of the Diet. The Diet members are called the Group of Twenty-two. They are professional politicians who have a special interest in foreign affairs. They belong to various political parties and have a variety of opinions. Talking with them through an interpreter, I was not sidetracked on minor technical questions of no real importance. If I had been talking with a similar group of American politicians, without the benefit of an interpreter, we would probably have spent the time arguing about technical details of MX and Pershing 2 missiles or about technical details of the Star Wars enterprise. I was happy to see that the Japanese politicians are interested in more basic and more important questions. What kind of military forces should Japan possess? To what extent should Japan rely on the United States for defense? These questions were discussed dispassionately and with a long-range historical perspective. I found the slow and scholarly discussion a refreshing change from the ephemeral style of political debate to which I am accustomed in America.

The Japanese politicians are acutely aware of the paradoxes of their status as allies of the United States. For Japan, to be pacifist means to rely for defense on the American nuclear umbrella. To be anti-nuclear means to take seriously the building up of independent Japanese armed forces. The propaganda of the peace movement urges Japan to be both pacifist and anti-nuclear. But the politicians know that the choice facing Japan in the real world is not so simple. The real choice is between nuclear pacifism and some degree of non-nuclear militarism. I was pleased to find the Group of Twenty-two clear-headed on this subject. They understand that each step going in the direction toward the abolition of nuclear weapons will give Japan greater independence and greater need for self-defense.

So far as I could tell, the Japanese politicians that I met were agreed on two essential points. First, the historic Japanese policy of not possessing, not producing, and not permitting nuclear weapons on Japanese territory should be maintained. Second, the Soviet occupation of Japanese territory in the northern islands since 1945 should not be condoned. I was surprised to find the desire for the recovery of the northern territories more acutely alive in Japan than the desire in Germany for the recovery of the corresponding territories in East Prussia and Silesia. From the Soviet point of view, the occupied Japanese islands are of far smaller political and strategic importance than the former German territories. The question then naturally arises, whether an Austrianization treaty might one day be negotiated between the Soviet Union and Japan, with Japan abandoning all vestiges of military alliance with the United States in return for a Soviet withdrawal from the northern islands.

I discussed this possibility with the Japanese politicians and they admitted that the idea was not unfamiliar to them. Some of them said they had tried to raise such questions in conversations with Soviet representatives. The Soviet response gave them no encouragement. Nevertheless, it would not be surprising if a deal of this kind between the Soviet Union and Japan should ultimately be worked out. In many ways, the Austrianization of Japan would be a less problematical venture than the Austrianization of Germany. In the case of Japan, there is an additional reason why an Austrianization treaty might seem advantageous to the Soviet Union. In the long run, the Soviet Union may be more concerned to prevent a military alliance between Japan and China than between Japan and the United States. For Japan, an Austrianization treaty would bring a number of lasting benefits. It would put to rest all doubts of Japan's non-nuclear status, it would lessen the risk of Japan becoming involved in American quarrels with third parties, and it would bring back the northern islands.

A visit to Hiroshima brings many other insights into Japanese life, besides the opportunity to pay formal respect to the dead. The living population of Hiroshima does not concern itself much with the dead. For the majority of the inhabitants today, the disaster of 1945 is ancient history, the Peace Museum is only for tourists. One of the loveliest Japanese parks is the Shukkei-en garden, hidden in the center of Hiroshima about one kilometer from ground zero. It has no connection with the larger and less beautiful Peace Park which was established after 1945. The Shukkei-en garden was there before the bomb and is still there today, a microcosm of Japan with little mountains, lakes, islands, bridges and shady walks where schoolchildren and pensioners take their morning stroll. A pilgrim coming to Hiroshima for the first time comes with feelings of tension, with thoughts of tragedy and doom. After two days in the city the tension is gone, the tragedy is a distant memory. The quietness of the Shukkei-en garden hushes the noise of the Peace Park.

Hiroshima is more relaxed than other Japanese cities. One of my young physicist friends told me he came to live in Hiroshima because it is a good place to bring up children. It has good swimming, good camping, good weather for outdoor living. It is the San Diego of Japan, a city of easy-going people enjoying a mild climate beside a sea of incomparable beauty. In the sea not far from the city there is an island called Miyajima where deer roam freely and groups of schoolchildren go camping. On the island there is a mountain with a cable car running up to the top. Inside the cable car there is a notice which sums up the spirit of Hiroshima insofar as a two-day pilgrim can grasp it. It says: "NOTICE. In case of emergency stop, wait for a moment quietly, please."

What does one learn from a visit to Hiroshima? The most important impressions one brings home are a renewed sense of the irrelevance of nuclear weapons and a renewed respect for the diversity of human nature. For forty years the world has been trying to fit Hiroshima into a preconceived role, to make Hiroshima remain forever an image of pity and horror, a static symbol without a past and without a future. But Hiroshima does not intend to become trapped into the role of symbol. For Hiroshima, the past and the future are more important than the accidental disaster that happened in between. The bomb begins to appear more and more as an irrelevant interruption in the city's life. Perhaps the reason why the city recovered from nuclear devastation with such astonishing resilience is that the survivors of the disaster in 1945 were guided by the same spirit which appears in the notice in the Miyajima cable car: "In case of emergency stop, wait for a moment quietly, please."

One of the qualities of Japanese life which immediately impresses a tourist is the habit of doing everything beautifully. Children out for a school excursion on the Miyajima ferry are beautifully dressed. Tea in the physics institute of the university is beautifully served. Tristan, the big particle accelerator in the national high-energy laboratory at Tsukuba, is beautifully sculptured. In the cheap shops that abound in the back streets of Tokyo, we wandered for hours without finding anything ugly. In the cheapest sushi bar the fish is beautifully arranged on the tray. I begin to see a connection between this pervasive pursuit of beauty and the Japanese attack on Pearl Harbor in 1941. In Japan, everything must be done beautifully, whether or not it makes sense. Many things which do not make sense are done because they are beautiful. The Pearl Harbor operation was a military work of art, a dramatic exercise in the elegant use of limited military resources. From a strategic point of view, it made no sense. Perhaps that is why it was done, because it was beautiful. In Japan, whether it makes sense is not the most important question.

Every country is unique, but Japan is more unique than the others. The Japanese are determined to preserve their uniquely illogical Kanji script, to keep alive the art of calligraphy in the age of computers. The Japanese passion for mechanical perfection has enabled Japan to beat Switzerland at the Swiss game of watchmaking and to beat America at the American game of carmaking. And still, as the political leaders of Japan well know, to be unique is not enough. Japan, with all its uniqueness, still has to find its place in the international order which we are trying to build.

Swiss armed neutrality provided the model for the neutralization of Austria and can perhaps provide a model for the political future of Japan. If there is any country in the world that resembles Japan in spirit, it is Switzerland. Switzerland too is obsessed with its own uniqueness. Switzerland too has grown rich by hard work and frugal living. Switzerland too is driven by a compulsion to do things beautifully, whether they make sense or not. One of my vivid memories of Switzerland is of an evening spent in the pharmacy in the town of Pontresina, high in the Engadin Valley. The pharmacist was a captain in the Swiss Army and was showing us films of his military exploits. His greatest achievement was to take an entire battalion of 250 mountain soldiers with skis and full equipment to the 4,000-meter summit of Piz Palü and down again without losing a man. He also had expert cameramen going to the summit with the soldiers to record the expedition for posterity. The whole operation was a magnificent exercise in traffic control. There were scenes of spectacular beauty at the summit and on the way down when the soldiers could finally put on their skis and schuss down the mountainside onto the glacier. Like the Pearl Harbor operation, it made no strategic sense but it was a great work of art. Fortunately our good pharmacist Captain Golay, unlike his opposite number Admiral Yamamoto in Japan, was planning his operations within a political tradition that allows Swiss soldiers to fight only in defense of Swiss territory. Switzerland has been an oasis of peace and sanity in the center of warring Europe for 170 years, from the Congress of Vienna until today. Why should not Japan become in international law, as it already is in commerce, the Switzerland of Asia?

Both Swiss and Japanese have a genius for making complicated machinery work. When I went with my wife recently to New York to buy a household knitting machine, the two available models came from Switzerland and from Japan. Both Swiss and Japanese rejoice in a language which keeps foreigners at a distance. The secretary of the physics institute in Kyoto told me that a certain foreign physicist visiting the institute was making everybody uncomfortable because he knows Japanese too well. I have seen the same uneasiness aroused in Zürich by a foreigner who is too serious in his study of Swiss-German. A foreigner speaking Schwyzerdütsch is, like a foreigner speaking Japanese, an invasion of privacy. Japan and Switzerland have also a third common characteristic. They both were once aggressive military powers and have a long tradition of soldiering. The Swiss military tradition finds expression in the citizen army, one of the most cohesive social institutions in the world. The Japanese military tradition has been in abeyance since 1945 but we must expect it to reappear in one form or another. The glorious victories of the 1904–05 war against Russia have not been forgotten. If ever the tradition of the Samurai should be revived in Japan, what form could it take that would be less dangerous to the peace of the world than an armed neutrality following the Swiss example?

The storming of Port Arthur has a place in the consciousness of Japan like the Battle of Waterloo in the consciousness of England. It was the classic decisive victory. Port Arthur was in 1904 the main Russian naval base in the Pacific. The storming of it was the bloodiest action of the Russo-Japanese War, costing the Japanese Army more than 30,000 dead. Dominating the approach to the base from the north was a hill, 203 meters high, which the Russians had strongly fortified. The hill had to be taken before the storming of Port Arthur could begin. The details of these events are familiar to every Japanese schoolchild.

One of my Japanese colleagues in Tokyo is director of the Institute of Space and Astronautical Science, which builds and launches unmanned scientific satellites. He wrote for the amusement of his friends a story with the title "An Old Soldier's Monologue." It is an informal account of the launch of the first Japanese X-ray astronomy satellite, written in a mixture of English and Japanese. The launch occurred on February 21, 1979. It was, like most pioneering scientific enterprises, a comedy of errors. The scientists monitoring the satellite's transmissions were unable for six weeks to decide whether the satellite was alive or dead. I quote now the director's account of what happened next. "On April 9, Murakami-san sent a Fax message saying, 'I found a big burst' in a format of military telegram. I immediately phoned and asked if he could see the modulation. He said, 'I clearly see it.' At the moment I recalled a famous scene during the Russo-Japanese War. Immediately after the long-awaited capture of the 203-meter hill on December 5, 1904, General Gentaro Kodama asked an artillery officer who ran up to the summit the most crucial question. 'Can you see Port Arthur?' Answer was, 'Yes, I can see it clearly.' I was really recalling that conversation. Childish?" With that question "Childish?" the director's story ends. The victory of the X-ray satellite Hakucho, seventy-five years after the victory at Port Arthur, cost nobody's blood. It was, of course, only a joke when the director of the Institute of Space and Astronautical Science imagined himself in the role of General Kodama seizing the famous 203-meter hill. Or was it only a joke? Even the director himself is not sure.

One thing at least is certain. It was a great victory for mankind when the tremendous talents and organizational skills of the Japanese people were deflected from the pursuit of military glory into peaceful commerce and science. Perhaps the peace activists of Hiroshima are right when they say that Japan has a unique mission to lead mankind into a future without nuclear weapons. To face the uncertainties of a non-nuclear world, mankind will need strong nerves and a quality which the Finns call *Sisu,* which may be translated into English as spiritual toughness. Spiritual toughness is a quality which Japan shares with Finland and with Switzerland. Early in the nuclear era, when America was armed with nuclear weapons and the Soviet Union was not, Joseph Stalin gave an accurate assessment of the utility of nuclear weapons. "The nuclear weapon," said Stalin, "is something to frighten people with weak nerves." Stalin did not suffer from weak nerves, and therefore he was not frightened. Unfortunately, his nerves were not strong enough for him to forego nuclear weapons altogether. While refusing to be intimidated by American bombs, he pushed the development of Soviet bombs as hard as he could.

The history of the last forty years has proved that neither the United States nor the Soviet Union had the spiritual toughness which is required to break loose from our addiction to nuclear weapons and move toward a non-nuclear international order. We have been afraid even to think seriously about dealing with nuclear weapons as the Japanese authorities dealt with guns in the seventeenth century. Japan set the world a good example then and is again setting the world a good example now. Seventeenth-century Europe, alas, did not follow the Japanese example of giving up the gun. Perhaps twenty-first-century Europe may be wiser. Perhaps we may this time be wise enough to follow the Japanese example.

15

NUCLEAR WINTER

"Nuclear winter" is a scientific theory which tries to describe the worldwide meteorological and ecological consequences of a nuclear war. The main result of the theory is that the aftereffects of nuclear war may be as lethal for people and plants and animals in countries remote from the war as for the populations of the belligerent countries. The phrase "nuclear winter" stands for a combination of radioactive and chemical pollution of the atmosphere with long-continued darkness and cold, destroying impartially the innocent and the guilty, the rain forest and the cornfield, the tiger and the rose.

When Carl Sagan and his colleagues began in 1983 to bring the possibilities of nuclear winter dramatically to the attention of the public, they put professional scientists like me into an awkward position. On the one hand, the professional duty of a scientist confronted with a new and exciting theory is to try to prove it wrong. That is the way science works. That is the way science stays honest. Every new theory has to fight for its existence against intense and often bitter criticism. Most new theories turn out to be wrong, and the criticism is absolutely necessary to clear them away and make room for better theories. The rare theory which survives the criticism is strengthened and improved by it, and then becomes gradually incorporated into the growing body of scientific knowledge. So, when the theory of nuclear winter appeared on the scene, my instinctive reaction as a scientist was to be skeptical, to look for the weak points and try to prove the theory wrong. That is the normal reaction to a new theory. We try to rip it apart as rapidly as possible.

On the other hand, nuclear winter is not just a theory. It is also a political statement with profound moral implications. If people believe that our nuclear weapons endanger not only our own existence and the existence of our enemies but also the existence of human societies all over the planet, this belief will have practical consequences. It will lend powerful support to those voices in all countries who oppose nuclear weapon deployments. It will increase the influence of those who consider nuclear weapons to be an abomination and demand radical changes in present policies. So my instinct as a scientist comes into sharp conflict with my instinct as a human being. As a scientist I want to rip the theory of nuclear winter apart, but as a human being I want to believe it. This is one of the rare instances of a genuine conflict between the demands of science and the demands of humanity. As a scientist, I judge the nuclear winter theory to be a sloppy piece of work, full of gaps and unjustified assumptions. As a human being, I hope fervently that it is right. Here is a real and uncomfortable dilemma. What does a scientist do when science and humanity pull in opposite directions?

There are three possible responses to the dilemma. One is to say, I am a scientist second and a human being first, so I will forget my scientific misgivings and jump onto the nuclear winter bandwagon. I do not blame those of my scientific colleagues who have taken this course. Survival is after all more important than technical accuracy. The second response is to say with the poet Robinson Jeffers:

> To seek the truth is better than good works, better
>   than survival,
> Holier than innocence and higher than love,

and since I am a scientist dedicated to truth, I will criticize nuclear winter as harshly as I would criticize any other half-baked scientific theory.

Few of my scientific colleagues have had the courage to make this second response. I do not blame those who have muffled their criticisms. Nobody wants to be put publicly in the position of saying that, after all, nuclear war may not be so bad. It is difficult to criticize nuclear winter publicly without giving an appearance of moral insensitivity. The majority of scientists who have doubts about nuclear winter keep their doubts private. This is the third response to the dilemma, to say, it will not do us any good in the long run to believe a wrong theory, but it will not do us any good in the short run to attack it publicly, so let us keep silent and reserve judgment until the facts become clear. I myself have chosen the third response. It is an unheroic compromise, but I prefer it to either of the simple alternatives. The dilemma is similar to the dilemmas which occur frequently in personal and family life, when the demands of honesty and of friendship pull in opposite directions. It is good to be honest but it is often better to remain silent.

There are political as well as scientific reasons for preferring a cautious response to nuclear winter. From a political point of view, it is dangerous to base a public campaign against nuclear weapons upon a technical argument which may later turn out to be wrong. The nuclear winter argument could backfire badly against the anti-nuclear movement if technical weaknesses are found in it. For a backfire effect to occur, it is not necessary that the nuclear winter theory be proved flatly wrong. It could also happen that the theory is proved to be right but that rather simple changes in weapon deployments and targeting rules will be sufficient to make the major effects of nuclear winter disappear. The opposition to nuclear weapons would then suffer a serious setback if it had emphasized nuclear winter too much in stating its case. The case against nuclear weapons is surely strong enough, the effects of nuclear war are catastrophic enough, the massacre of innocent people is abhorrent enough, so that a political movement against nuclear weapons need not depend on the technical details of nuclear winter for its justification.

There is perhaps a historical analogy between Linus Pauling in the 1950s and Carl Sagan in the 1980s. Linus Pauling was the great crusader against bomb testing in the 1950s. He richly deserved the Nobel Peace Prize which he won for his efforts. He succeeded in mobilizing public opinion all over the world against the massive hydrogen-bomb tests which were then occurring regularly in the Pacific and in the Arctic. Unfortunately, he used as his main argument against bomb testing the medical and genetic damage caused by radioactive fallout resulting from the tests. The public outrage which he mobilized was directed primarily against the fallout and not against the weapons. As a result, the outrage collapsed as soon as the Limited Test Ban Treaty of 1963 was signed and the bomb testing was hidden underground. The Limited Treaty stopped the fallout but did not stop the nuclear arms race. Pauling himself had campaigned for a complete Test Ban which would stop all tests, but his overemphasis of the fallout issue prevented him from reaching his objective. Since the public was concerned about fallout rather than about weapons, the public was satisfied with the Limited Test Ban, and the development of weapons continued almost unhindered.

It is easy to imagine a similar sequence of events arising out of Carl Sagan's nuclear winter campaign. Carl Sagan deserves enormous credit for raising the nuclear winter question and forcing us to think about it. He is as much a hero of our time as Pauling was the hero of the fifties. The nuclear winter question, like the fallout question in the fifties, gives hope and strength to everybody who is fighting the political battle against nuclear weapons. Nuclear winter is a simple and powerful idea around which opposition to nuclear weapons can crystallize. Unfortunately, the nuclear weapons establishment may find the equivalent of a Limited Test Ban Treaty, a simple technical fix which would remove the danger of nuclear winter without removing the danger of nuclear war. I am not saying that this is easy, but we do not yet know enough to say that it is impossible. With this possibility in mind, I give my blessing to Carl Sagan's campaign but continue to feel some anxiety about his tactics. The wave of moral outrage which Sagan has created must be directed against the evil of nuclear war itself and not merely against its consequences.

I do not wish here to get into a technical argument about the details of nuclear winter. I will merely summarize my own struggles with the technical issues. I spent a few weeks in 1985 trying to make nuclear winter go away. The phrase "go away" here is used in the sense customary among scientists. To destroy a new theory, you try to find a simple situation where the theory predicts that something happens and you can prove that the same something does not happen. Then you say that the thing predicted has gone away and the theory is demolished. I found, after about two weeks of work, that I could not make nuclear winter go away. That is to say, I could not prove the theory wrong. As a result, I now understand the theory better and believe it much more than I did when it was first announced. But I do not yet believe it a hundred percent. Our technical understanding of it is still rudimentary. It is still possible that when we understand it better we will find that it has gone away.

When scientists are arguing about a complicated and controversial problem, it often happens that their technical judgments are influenced more by personal experience than by objective calculation. The arguments about nuclear winter show clearly the influence of personal background. For example, Carl Sagan believes strongly in the reality of nuclear winter whereas I am still skeptical. We both use the same mathematics and both work with the same laws of physics. Why then do we reach different conclusions? Perhaps the difference in our conclusions results from the fact that Carl

Sagan spent a large fraction of his life studying Mars while I spent a part of my life living in London. Carl Sagan's intuition about nuclear winter is based on his experience of Martian dust storms. My intuition about nuclear winter is based on my experience of old-fashioned London fog. The main difference between a Martian dust storm and a London fog is that Mars is dry while London is wet.

Carl Sagan has vivid memories of the great dust storm which was raging all over the planet when the Mariner 9 spacecraft came to make the first detailed observations of Mars in 1971. He remembers the temperature on the Martian ground falling as the dust clouds obscured the Sun. I have vivid memories of an orchestral concert in the Albert Hall in London, when I sat in the front row of the circle and heard the music emerging mysteriously from an invisible orchestra behind a wall of fog. I remember that on this and other occasions when London was enveloped in fog, the temperature on the ground under the fog was warm. Carl Sagan's image of nuclear winter is based on the optical properties of dry dust; my image of nuclear winter is based on the optical properties of wet soot. Dry dust or dry soot acts as an optical filter, blocking sunlight from reaching the ground but allowing heat radiation from the ground to escape into space. Wet dust or wet soot acts as a blanket, blocking all radiation from carrying energy either upwards or downwards.

If the atmosphere after a nuclear war is filled with dry soot, the temperature on the ground will fall and the Earth will experience nuclear winter. If the atmosphere is filled with wet soot, the temperature on the ground will stay roughly constant as it used to do under a London fog. The severity of a nuclear winter depends on whether the soot-laden atmosphere is predominantly dry or predominantly wet. In nuclear winter as in the meteorology of normal terrestrial weather, water in its multifarious forms of vapor, cloud, rain and snow plays a dominant role. We live on a water-dominated planet. And the prediction of weather patterns is still, even with all the help we can obtain from large-scale computer simulations, more an art than a science. Meteorologists still must use intuition and experience to make their predictions. Carl Sagan's intuition has the Earth after a nuclear war resembling Mars under a dust storm; my intuition has the Earth resembling London under a fog, and the computers are not clever enough to tell who is right.

These technical arguments about the meteorological consequences of nuclear war should not be taken too seriously. The essential fact, which is true independently of the meteorological uncertainties, is that the consequences of a nuclear war are incalculable and unpredictable. Nuclear winter has only added an additional dimension to the incalculability and unpredictability of war. Nevertheless, the new dimension of unpredictability is important. No matter how the technical arguments may be settled or remain unsettled, nuclear winter stands as a symbol of the vulnerability of mother Earth to the violence of her human children. Nuclear winter is not primarily a technical problem; it is much more a moral and political problem. It forces us to ask fundamental questions: whether the benefits which we derive from the possession of nuclear weapons are in any way commensurate with the risks; whether the risk of irreparable damage to the fabric of life on Earth can in any way be morally justified.

The nuclear winter issue, like the radioactive fallout issue in the fifties, has the great virtue of being global. It compels the nuclear powers to accept some moral responsibility for the damage we may do to mankind and to our planet as a whole. But our awareness of this responsibility should not depend on whether the technical details of nuclear winter theory turn out to be right or wrong. Quite apart from nuclear winter, a nuclear war would be an unconscionable crime against humanity. The worldwide spread of famine and disease resulting from economic and social disruption may be as murderous as the physical effects of warfare. The discovery of nuclear winter has given a new starkness to our responsibility for the survival of mankind. Even without nuclear winter, we are still responsible.

British and American thinking about nuclear weapons is strongly influenced by two popular myths surrounding their origin. One myth says that nuclear weapons were decisive in winning World War II. The second myth says that if Hitler had got nuclear weapons first, he could have used them to conquer the world. Both myths were believed by the scientists and statesmen who built the first nuclear weapons. They are still believed by most Americans today. Since we cannot explore the might-have-beens of history, we cannot know for sure whether the myths are true. All that we can say for sure is that these myths have created the unexamined background for later American doctrine concerning nuclear weapons. Nuclear weapons are seen as militarily decisive (because we won) and historically justified (because Hitler might have conquered the world).

I believe that both myths are false. Of course I cannot prove it. But it is important to look at the myths with a skeptical eye, and to consider how different our view of nuclear weapons might have been if Hitler had in fact got them first. Suppose that Britons and Americans had neglected to push nuclear weaponry seriously and that the Germans had pushed as hard as possible. Hitler might have had a bomb by 1943 at the earliest, and perhaps a few tens of bombs by 1945. What difference would it have made? London and Moscow would no doubt have shared the fate of Hamburg and Dresden. Perhaps a few square miles of New York would have been demolished. A lot of people would have been killed. But it seems highly unlikely that the arrival of Russian soldiers in Berlin and of American soldiers in Tokyo would have been substantially delayed. We saw the actual effects of heavy bombing in England and in Germany in World War II. The effects were to make the countries more united, to remove the barriers which usually separate civilians from front-line soldiers, to create a spirit of toughness and resilience in the entire population. The same effects were seen twenty-five years later in Vietnam. If Hitler had had nuclear bombs, their use would have neither changed the grand strategy of the war nor lessened our determination to fight it to a finish. What would have been changed is our postwar perception of nuclear weapons. We would have seen nuclear weapons forever afterwards as contemptible, used by an evil man for evil purposes and failing to give him victory. The myth surrounding nuclear weapons would have been a myth of contempt and failure, not a myth of pride and success.

It is important for Britons and Americans to go through the mental exercise of looking at nuclear weapons as if they had been Hitler's weapons rather than ours, because this exercise enables us to come closer to seeing nuclear weapons as they are seen by Soviet citizens. The dropping of the bombs in 1945 is seen by many Russians as an act of intimidation directed against them. To understand Russian strategy and Russian diplomacy, it is necessary for us to distance ourselves from our own myths and to enter into theirs. An understanding of Soviet viewpoints is an essential first step toward amelioration of the danger in which the world now stands.

If a political arrangement is to be durable, it must pay attention both to technical facts and to ethical principles. Technology without morality is barbarous; morality without technology is impotent. But in the public discussion of nuclear policies in the United States, technology has usually been overemphasized and morality neglected. It is time for us now to redress the balance, to think more about moral principles and less about technical details. The roots of our nuclear madness lie in moral failures rather than in technical mistakes. Our thinking is permeated by our historical myths. We tend to accept without serious question the idea that nuclear weapons are militarily decisive. We usually equate military effectiveness with destructive power. We rarely examine critically the military purpose of nuclear weapons or the possible missions for which they might be used. The case for the feasibility of abolishing nuclear weapons would be stronger if we treated them with less respect. The hope of successful abolition becomes more realistic if it is understood that nuclear weapons are absurd rather than omnipotent.

As an example of our overrating of nuclear weapons, consider the alleged restraint of the United States during the years when we held a nuclear monopoly. We often hear claims that the United States behaved with peculiar virtue in the postwar years because we did not either annihilate the Soviet Union or use nuclear blackmail to force the Soviet Union out of Eastern Europe. The idea that we could have annihilated the Soviet Union with our meager supply of bombs is totally unreal, and there was never a time when nuclear blackmail would have had much chance of success.

American political leaders often talk as if the primary purpose of nuclear weapons were to massacre people without much regard for military utility. In fact the great majority of nuclear weapons are deployed for a precisely opposite purpose, to destroy opposing military forces while massacring as few people as possible. The technical characteristics of the weapons are wholly unsuited to their militarily desirable use. By misconceiving their purpose, we misconceive their effectiveness. As a corollary of this misconception, we denigrate the effectiveness of defense against nuclear weapons. If the purpose of nuclear weapons were simple massacre, a defense would indeed be impossible. But if, as is actually the case, nuclear weapons are deployed with missions for which they are unsuited or only marginally effective, an imperfect defense may be able to tip the balance against them.

The most serious weakness of American strategic thinking is lack of respect for Soviet points of view. In an odd way, Soviet nuclear doctrines come closer than ours to the point of view of the Catholic bishops. Soviet doctrine, like the Catholic bishops, forbids deliberate targeting of civilian populations, forbids the first use of nuclear weapons, and rejects deterrence as an ultimate strategic goal. The Soviet Union offered to negotiate an abolition treaty in 1946 and the United States rejected it. Perhaps the best way to achieve an abolition treaty would be to pick up the negotiation of the Soviet proposal where we left it in 1946. Secretary Gorbachev in his recent proposals for nuclear disarmament reminded us of the Soviet 1946 offer. We paid little attention to the offer, either in 1946 or later, but the Russians still remember it.

Americans tend to think of nuclear weapons as an invincible force of which we should be mortally afraid. Stalin knew better. If we are to succeed in abolishing nuclear weapons, it is not enough to be mortally afraid. We shall have a better chance if we understand that nuclear weapons are useless and dangerous toys which we are free to discard if our nerves are strong.

Everybody who travels around the world as I do, giving lectures about nuclear weapons and nuclear war, must sometimes feel a stirring of moral qualms. What right have I to enjoy public acclaim and hospitality, while I talk cheerfully about weapons which bring agonizing death to millions? These qualms were expressed poignantly by Mark Kaminsky, a young American poet who wrote *The Road from Hiroshima.* Kaminsky's book is a cycle of poems telling of the bombing of Hiroshima, using so far as possible the actual words of the victims. Kaminsky has been reciting his poems in public and drawing big audiences. At the end of his book there is a poem, "The Witness," expressing the qualms which I also have reason to share:

> As I step off the platform
>
> After reciting my poems
> I am abashed.
> I feel more ashamed than ever, facing
> The tears and gratitude I evoke
> And my immense hunger for both.
>
> Suddenly,
> I can't tell the difference between being
> A profiteer on the spiritual black market
> And a prophet
> Who must tear everyone's heart to shreds.

As I wrote,
I felt possessed by the dead calling for peace.

Now I wonder,
Is it so laudable
To spend my days summoning images of nuclear war?

Yet how can I give up this fire,
Without betraying myself and all that I love?

16

THE TWENTY-FIRST CENTURY

Technology is a gift of God. After the gift of life it is perhaps the greatest of God's gifts. It is the mother of civilizations, of arts and of sciences. Nuclear weapons are a part of technology, but technology has outgrown nuclear weapons just as it has outgrown other less crude instruments of power. Technology continues to grow and to liberate mankind from the constraints of the past. Compared with the revolutions which technology is bringing to people and institutions all over the world, our quarrels with the Russians are small motes in the eye of history.

The most revolutionary aspect of technology is its mobility. Anybody can learn it. It jumps easily over barriers of race and language. And its mobility is still increasing. The new technology of microchips and computer software is learned much faster than the old technology of coal and iron. It took three generations of misery for the older industrial countries to master the technology of coal and iron. The new industrial countries of East Asia, South Korea and Singapore and Taiwan, mastered the new technology and made the jump from poverty to wealth in a single generation. That is the reason why I call the new technology a technology of hope. It offers to the poor of the Earth a short-cut to wealth, a way of getting rich by cleverness rather than by back-breaking labor. The essential component of the new technology is information. Information travels light. Unlike coal and iron, it is available wherever there are people with brains to make use of it. Not only in East Asia but all over the planet, technology and the information on which it depends can be effective instruments for achieving a more just distribution of wealth among the nations of mankind. Without the hope of economic justice, mankind cannot realistically hope for lasting peace. If we view the world with a certain largeness of view, we see technology as the gift of God which may make it possible for us to live at peace with our neighbors on this crowded planet.

Such a largeness of view is conspicuous by its absence in the thinking of the Reagan administration. I dislike many things which this administration has done and said, but I dislike most of all the mean-spirited attempts to stop the export of technology and hamper the spread of information. These attempts reveal a mentality which is incompatible with any decent respect for the opinions of mankind. The idea that the United States should try to keep the Soviet Union in a state of technological backwardness excludes the possibility of comprehensive arms-control agreements; the Soviet Union will not negotiate upon any terms other than equality. The idea that the United States can play Nanny to the rest of the world and constrain the flow of technological goodies to reward our friends and punish our enemies is a puerile delusion. Technology is God's gift to all nations alike. The rest of the world will quickly learn whatever we attempt to keep hidden. And we will quickly lose the international goodwill which a more generous attitude has earned us in the past. If we are to lead the world toward a hopeful future, we must understand that technology is a part of the planetary environment, to be shared like air and water with the rest of mankind. To try to monopolize technology is as stupid as trying to monopolize air.

Technology as a liberating force in human affairs is more important than weapons. And that is why scientists speak about international political problems with an authority which goes far beyond their competence as bomb-builders. Forty years ago, scientists became suddenly influential in political life because they were the only people who knew how to make bombs. Today we can claim political influence for a better reason. We claim influence because we have practical experience in operating a genuinely international enterprise. We have friends and colleagues, people we know how to deal with, in the Soviet Union and in the People's Republic of China. We know what it takes to collaborate on a practical level with Soviet scientists, the bureaucratic obstacles that have to be overcome, the possibilities and limitations of personal contact. We know what it takes to operate an astronomical observatory in Chile, to launch an X-ray satellite from Tanzania, and to organize the eradication of the smallpox virus from its last stronghold in Ethiopia. Unlike our political leaders, we have first-hand knowledge of a business which is not merely multinational but in its nature international. We know how difficult it is to get a piece of apparatus to work in the Soviet Union or in China, but we also know how with patience it can be done. As scientists we work every day in an international community. That is why we are not afraid of the technical difficulties of arms control. That is why we are appalled by the narrow-mindedness and ignorance of our political leaders. And that is why we are not shy to raise our voices, to teach mankind the hopeful lessons that we have learned from the practice of our trade.

It usually takes fifty to a hundred years for fundamental scientific discoveries to become embodied in technological applications on a large enough scale to have a serious impact on human life. One often hears it said that technological revolutions today occur more rapidly than they did in the past. But the apparent acceleration of technological change is probably an illusion caused by perspective. Recent events are seen in greater detail than historical events of a century ago, and the loss of detail makes the more remote technological changes appear to proceed more slowly. In reality, the time elapsed between Maxwell's equations and the large-scale electrification of cities was no longer than the time between Thompson's discovery of the electron and the worldwide spread of television, or between Pasteur's discovery of microbes and the general availability of antibiotics. In spite of the hustle and bustle of modern life, it still takes two or three generations to convert a new scientific idea into a major social revolution.

If it is true that the interval between discovery and large-scale application is still of the order of seventy years, this means that we should be able to foresee with some reliability the main technological changes that are likely to occur up to the middle of the twenty-first century. Until about the year 2050, large-scale technologies will be growing out of discoveries which have already been made. Only after 2050 are we likely to encounter technologies based on principles unknown to our contemporary science.

Here are my guesses for the dominant new technologies of the next seventy years. I look at contemporary science and see three main areas of existing knowledge not yet fully exploited. These are the same three areas which I invoked for the design of the Astrochicken spacecraft in Chapter 10. The first is molecular biology, the science of genetics and cellular physiology at the molecular level. The second is neurophysiology, the science of complex information-processing networks and brains. The third is space physics, the exploration of the solar system and the physical environment of the Earth. Each of these areas of science is likely to give rise to a profound revolution in technology. The names of the new technologies are genetic engineering, artificial intelligence and space colonization. This short list is not complete. No doubt there will be other innovations of equal importance. Whatever else may happen, these three technological revolutions will be changing the conditions of human life during the coming century. I will say a few words about each of them in turn.

Genetic engineering is already established as a tool of manufacture in the pharmaceutical industry. Bacteria can be infected with alien genes and cloned to produce in quantity the proteins which the alien genes specify. But the quantities that can be produced in this way are at present small. Genetic engineering makes economic sense today only for producing drugs which can be sold at a high unit price. Genetic engineering does not yet begin to compete with conventional industrial processes for the mass production of common chemicals. The fundamental limitation of genetic engineering as it now exists is the limitation of through-put. A genetically engineered bacterium in a tank produces about as much material in a day as a conventional combustion reactor in the same tank would produce in a second. Biological reactions are slow and require large volumes to produce substantial through-put of products. For this reason, genetic engineering will not replace conventional chemistry so long as the genetically engineered creatures are confined in tanks and retorts.

But why should genetically engineered production processes be confined in tanks? One reason for confinement is concern for environmental safety. Regulations in most countries forbid us to let genetically engineered creatures loose in the open air. It is reasonable to be cautious in relaxing regulations. Fears of genetically engineered monsters overrunning the Earth are often exaggerated, but the dangers may not be altogether imaginary. Newly engineered creatures must be studied and understood before they are released into the great outdoors. Still it seems likely that we shall learn in time to transfer genetic-engineering technology from the enclosed tank to the open field without serious danger. After all, farmers have been growing wheat in open fields for thousands of years, and wheat is also a product of human manipulation, just as artificial as genetically engineered *Escherichia coli*. Farmers long ago discovered that it is more profitable to grow wheat in open fields than in greenhouses. Genetic engineering will likewise become profitable for large-scale chemical production when the growing and harvesting of genetically engineered species can be moved outdoors. Chemical industry will then no longer be clearly distinguishable from agriculture. Crop plants will be engineered to produce food or to produce industrial chemicals according to demand.

Every technological revolution causes unplanned and unwelcome side effects. Genetic engineering will be no exception. One of the harmful side effects of genetic engineering might be the displacement of traditional agriculture by industrial crops. This effect would be a further extension of the displacement of subsistence farming by cash crops which happens today in developing countries. The disappearance of subsistence farming is deplorable for many reasons. It causes depopulation of the countryside and overgrowth of cities, it reduces the genetic diversity of our crop species, and it destroys the beauty of traditional rural landscapes. Already in 1924, J. B. S. Haldane saw what was coming and described it in his book *Daedalus*. Here is a passage quoted by Haldane from a term paper written by a fictitious undergraduate in the twenty-first century, summarizing the effects of genetic engineering in the twentieth:

> As a matter of fact it was not until 1940 that Selkovski invented the purple alga *Porphyrococcus fixator* which was to have so great an effect on the world's history. . . . *Porphyrococcus* is an enormously efficient nitrogen-fixer and will grow in almost any climate where there are water and traces of potash and phosphates in the soil, obtaining its nitrogen from the air. It has about the effect in four days that a crop of vetches would have had in a year. . . . The enormous fall in food prices and the ruin of purely agricultural states was of course one of the chief causes of the disastrous events of 1943 and 1944. The food glut was also greatly accentuated when in 1942 the Q strain of *Porphyrococcus* escaped into the sea and multiplied with enormous rapidity. When certain of the plankton organisms developed ferments capable of digesting it, the increase of the fish population of the seas was so great as to make fish the universal food that it is now. . . . It was of course as the result of its invasion by *Porphyrococcus* that the sea assumed the intense purple colour which seems so natural to us, but which so distressed the more aesthetically minded of our great grand-parents who witnessed the change. . . . I need not detail the work of Ferguson and Rahmatullah who in 1957 produced the li-

chen which has bound the drifting sand of the world's deserts, for it was merely a continuation of that of Selkovski, nor yet the story of how the agricultural countries dealt with their unemployment by huge socialistic windpower schemes. . . .

Haldane had his dates wrong by fifty years. He expected the genetic engineering revolution to come in the 1940s. In fact it is coming in the 1990s or later. But there is no doubt that it will come. Haldane also understood that it will not be an unmixed blessing. As he says at the end of his book:

> The scientific worker is brought up with the moral values of his neighbours. He is perhaps fortunate if he does not realize that it is his destiny to turn good into evil. An alteration in the scale of human power will render actions bad which were formerly good. Our increased knowledge of hygiene has transformed resignation and inaction in face of epidemic disease from a religious virtue to a justly punishable offence. We have improved our armaments, and patriotism, which was once a flame upon the altar, has become a world-devouring conflagration.

One of the benefits of the genetic engineering revolution will be to allow us to make great areas of the globe economically productive without destroying their natural ecology. Instead of destroying tropical forests to make room for agriculture, we could leave the forests in place while teaching the trees to synthesize a variety of useful chemicals. Huge areas of arid land could be made fruitful either for agriculture or for biochemical industry. There are no laws of physics and chemistry which say that potatoes cannot grow on trees or that diamonds cannot grow in a desert. Moreover, animals can be genetically engineered as well as plants. There are no laws of nature which say that only sheep can produce wool or that only bees can harvest honey. In the end, the genetic engineering revolution will act as a great equalizer, allowing rich and poor countries alike to make productive use of their land. A suitably engineered biological community will be able to produce almost any desired chemical from air, rock, water and sunshine. Ultimately even water may be unnecessary, since the driest desert air contains enough water vapor to sustain a biological community if the community is careful not to waste it. Fixing water from the air should be a simpler biochemical problem than fixing nitrogen, and many existing plants know how to fix nitrogen.

It is easy to talk in general terms about the effects of new technology, but difficult to assess the effects of particular technologies on particular places. Since I gave these lectures in Aberdeen, I consider Aberdeen as an example. Aberdeen is like many European cities, old and beautiful and adapting itself with some success to the demands of twentieth-century life. Who could have predicted a hundred years ago the effects which the automobile would produce on the demography and pattern of growth of Aberdeen, the dispersal of homes and the conversion of farmland into suburbs? Nobody can be wise enough or imaginative enough to predict the effects which the genetic engineering revolution will have on the landscape of Scotland. Genetic engineering is a tool which can be used wisely or unwisely. The effects will be good or bad according to our wisdom or unwisdom. If we are wise, using the tools of genetics with artistry and with respect for the indigenous ecology, the Scottish Highlands might be made rich and fertile without being made ugly.

The second technological revolution is artificial intelligence. This revolution has already begun with the rapid development and proliferation of computers. I see the effects of the revolution already in the sociology of the Institute in which I work at Princeton. Until recently, our visiting members were usually assigned to offices with two members to an office. The offices are rather small but there was room in each office for two desks and two chairs. The members were generally content to share offices. It gave them a chance to make friends and to be drawn into scientific collaborations. If the office-mate smoked too much or talked too much, rearrangements could be peacefully negotiated. But, alas, this happy sociological pattern is now no more. A few years ago the Institute decided that in order to remain competitive with other research institutions, we must provide our members with computer terminals. In each office now there is a computer terminal which has to sit on one of the two desks. There is no longer enough room for two humans to work comfortably side by side. Each office now houses one human and one terminal, and the surplus members are exiled to another building.

Computer terminals in offices and homes are only the beginning of artificial intelligence. Artificial intelligence is an enterprise with grander aims. In discussing the future of artificial intelligence, I shall be following the script written by my friend Sir James Lighthill fifteen years ago. In 1972 Lighthill surveyed the field of artificial intelligence in an official report commissioned by the British Science Research Council. He was asked to evaluate the work done in the United Kingdom up to 1972 and to estimate the prospects for further progress up to the year 2000. We are now halfway between 1972 and 2000. So far, the development of the field has conformed closely to Lighthill's predictions. I consider it likely that Lighthill's estimates will remain valid up to the year 2000. But since I am interested in a longer future, I shall depart from Lighthill's script when I offer my guesses concerning what may happen later.

Lighthill begins by dividing artificial intelligence into three areas which he calls A, B, and C. A stands for advanced automation, the objective being to replace human beings by machines for specific purposes, for example, industrial assembly, military reconnaissance or scientific analysis. A large body of work in category A is concerned with pattern recognition, with the programming of computers to read documents or to recognize spoken words. C stands for computer-based central-nervous-system research. The objective here is to understand the functioning of brains, either human or animal, using the computer as a tool to complement and interpret the facts of experimental neurophysiology. A more remote aim is to understand the architecture of the brain so completely that we can borrow the brain's architecture in building a new generation of computers. Finally, B stands for bridge, an area of work which aims to make contact between A and C, to make use of neurophysiological models in designing machines to perform practical tasks. The main activity in area B has been the building of robots. Lighthill's main conclusion is that while work in areas A and C is promising and worthy of support, area B is largely illusory. Both advanced automation and neurophysiology are real sciences with concrete achievements, but the bridge linking them together is nonexistent. Insofar as artificial intelligence claims to be the unifying bridge, artificial intelligence has no real existence.

In the United Kingdom, Lighthill's sweeping condemnation of area B had the effect of a self-fulfilling prophecy. Funding of efforts in area B was withdrawn, and areas A and C continued their independent development. But in other parts of the world, and in particular in the United States, the same decline of area B occurred without Lighthill's intervention. I conclude that Lighthill's diagnosis was accurate, that his harsh words about area B were well founded.

Here is Lighthill's famous caricature of area B:

> Most robots are designed to operate in a world as like as possible to the conventional child's world as seen by a man: they play games, they do puzzles, they build towers of bricks, they recognize pictures in drawing-books, "bear on rug with ball," although the rich emotional character of the child's world is totally absent. Builders of Robots can justly reply that while robots are still in their infancy they can mimic only pre-adult functions and a limited range of those at most, and that these will lead on to higher things. Nevertheless, the view to which this author has tentatively but perhaps quite wrongly come is that a relationship which may be called pseudomaternal comes into play between a Robot and its Builder.

As Lighthill predicted, the fruitful development of artificial intelligence during the last ten years has occurred in area

A and not in area B. The successful programs are utilitarian tools designed to perform specific tasks without any pretensions of intelligence. They are not supposed to understand what they are doing, nor to mimic the operations of a human intelligence. Their software incorporates large quantities of human knowledge, but this knowledge is supplied to them from the outside, not generated on the inside by any process of internal ratiocination. Artificial intelligence has been practically useful only when it abandoned the illusion of being intelligent.

What of the future beyond the year 2000? I agree with Lighthill in expecting advanced automation and neurophysiology to continue to develop as separate sciences. They still must grow within their separate domains before bridge building will be possible. But sooner or later the two areas are bound to come into contact. The time will come when brain-architecture in area C begins to be understood in detail and program-architecture in area A begins to acquire some of the sophistication of natural human language. At that stage the time will be ripe for building bridges, and the further progress of the two areas will be merged rather than separate. Machine builders will be able to incorporate the structures of neurophysiology into their designs, and neurophysiologists will be able to monitor neural processes with properly matched connections between brains and computers. When progress has reached this point, the grand claims of artificial intelligence, so prematurely made and so justly ridiculed, will at last be close to fulfillment. The building of truly intelligent machines will then be possible. The artificial intelligence revolution will be upon us in full force.

How long will this take to happen? My guess is about fifty years from now, some time between the years 2000 and 2050. I am old enough so that I do not need to worry about seeing my guess proved wrong. What will be the human consequences of artificial intelligence? To guess the consequences is even more hazardous than to guess the date of the revolution. I will say only that my view of the consequences is not apocalyptic. I do not see any real danger that human intelligence will be supplanted by artificial intelligence. Artificial intelligence will remain a tool under human control. To conclude my assessment of the future of artificial intelligence, I quote again from Lighthill:

> The intelligent problem-solving and eye-hand co-ordination and scene analysis capabilities that are much studied in category B represent only a small part of the features of the human central nervous system that give the human race its uniqueness. It is a truism that human beings who are very strong intellectually but weak in emotional drives and emotional relationships are singularly ineffective in the world at large. Valuable results flow from the integration of intellectual activity with the capacity to feel and to relate to other people. Until this integration happens, problem-solving is no good, because there is no way of seeing which are the right problems. The over-optimistic category-B-dominated view of artificial intelligence not only fails to take the first fence but ignores the rest of the steeplechase altogether.

My verdict agrees with Lighthill's. I believe that artificial intelligence will succeed in jumping the first fence before the year 2050, but that human intelligence is far ahead and will remain far ahead in the rest of the steeplechase, as far into the future as I can imagine. Man does not live by problem solving alone. Artificial intelligence will not only help us with solving problems, but will also give us freedom and leisure for exercising those human qualities which computers cannot touch.

The third technological revolution which I see coming is the expansion of life's habitat from Earth into the solar system and beyond. This revolution may take a little longer than the other two. Perhaps it may take as long as a hundred years from now. In charting a possible course for this revolution, I take as my guide Ben Finney, an anthropologist at the University of Hawaii who has made a detailed study of the Polynesian navigators and their voyages of colonization from island to island across the Pacific Ocean. The Polynesians did not travel alone but carried with them as many helpful plants and animals as possible. And we shall too. Only for us, bringing life for the first time into a lifeless wilderness, the bringing along of a sufficient variety of plants and animals will be even more essential.

Finney and his friend Eric Jones wrote an essay with the title *From Africa to the Stars,* surveying the course of human history in the past and the future. Jones is not an anthropologist. He is a space scientist at the Los Alamos National Laboratory in New Mexico. Finney is an expert on the past and Jones is an expert on the future. They condense the whole of human history into four big steps. Step 1 was taken about four million years ago in East Africa. It was the step from the trees to open grassland. The new skills required for the change were walking and carrying. Step 2 was the move out of the warm, sunny climate of Africa to the more varied and generally hostile habitats of the remaining continents, Asia and Europe and America and Australia. This step began about 1 million years ago. The new skills required for it were hunting, firemaking and probably speech. Step 3 was the move from land out onto the open sea. This step began three thousand years ago and was taken first by the Polynesians, with the Europeans following hard on their heels. The new skills required were shipbuilding, navigation and science.

Step 4 is the step from Earth to the stars. This step is beginning now and will occupy us for at least the next few hundred years. The new skills required are to some extent already in hand: rocketry, radio communication, observation and analysis of remote objects. But Step 4 is a bigger enterprise than our present-day space technology can handle. Step 4 is the permanent and irreversible expansion of life's habitat from Earth into the cosmos. It will require other new skills which are not yet in hand. It will require genetic engineering, and probably artificial intelligence too. Genetic engineering to allow colonies of plants and animals to put down roots, to grow and spread in alien environments. Artificial intelligence to allow machines to go out ahead of life and prepare the ground for life's settlement. This is not to say that Step 4 cannot begin until the genetic engineering and artificial intelligence revolutions are complete. Step 4 is already in progress in a preliminary and tentative fashion. We are already started on our way to the stars. But this step, like the first three steps from trees to grassland, from Africa to the world, from land to sea, will not be finished within a century.

In Chapter 6, I wrote in general terms about the adaptation of life to alien environments. I was thinking then about the very long run, about life considered as a universal phenomenon, extending over distances and times of cosmic dimensions. Here I am looking in greater detail at the nearer future, to see what life might do in our own little corner of the universe within the next few centuries.

Some of the first questions which come up in any practical discussion of space colonization are questions of economics. Suppose we go out and settle on a convenient asteroid with our little spaceship, what do we do when we get there? How do we make a living? What can we expect to export in order to pay for necessary imports? If space colonization makes any sense at all, these questions must have sensible answers. Unfortunately, we cannot hope to answer questions of economics until the asteroids have been explored. At present we know almost nothing about the chemical resources of asteroids and the physical conditions we shall find there. The most important of all resources is water, and the abundance of water is still unknown. No human instrument has ever touched an asteroid, or come close enough to make detailed observations. In this connection, it is interesting to compare the economics of asteroid settlement with the economics of early colonies in North America. The early American colonists knew almost as little about America as we know about asteroids, and their economic expectations almost always proved to be wrong. The first settlers in Virginia intended to find gold and instead grew rich by exporting tobacco. The Pilgrims in Massachusetts intended to live mainly by fishing and instead became farmers and fur traders. The most important prerequisite for economic survival is flexibility. Colonists should never believe economic forecasts and should be ready to switch to other means of livelihood when the forecasts are wrong.

In the circle of my own friends I have seen an example of a venture of colonization which may throw some light on the economic problems of asteroid settlers. My friends, a young man and his wife, established themselves on an uninhabited island in the North Pacific. They built themselves a comfortable house and had no difficulty in growing enough food for the table. The husband was a skilled blacksmith and built a sawmill and other useful pieces of machinery. In most respects the colony was economically self-sufficient. But there were a few essential items which they could not produce for themselves and needed to import. Diesel fuel was the most essential import. They needed to sail down to Vancouver once or twice a year with a large barrel and fill it up with diesel fuel for their engines. The question then arose, What was the most convenient cash crop which they could produce for export? They had to find a crop which could be easily grown on their island, easily transported in their boat, and easily sold for a high price in Vancouver. They were law-abiding citizens of Canada, and they had no wish to become involved with the smuggling of illegal drugs. What then is the most convenient, legal, high-value cash crop for a small island in the North Pacific to export? The answer to this question was not obvious. My friends only found the answer by accident after a number of false starts. The answer was, pedigree Rhodesian Ridgeback pups. The dogs were easy to breed on the island. They did not need to be fenced since there was no danger of miscegenation. They fed mostly on leftovers from the farm. And the pups could be sold to dog-fanciers in Vancouver for a couple of hundred dollars each.

The economic problems of asteroid colonists must be solved in a similar fashion, by finding cash crops which conveniently exploit local opportunities. Rhodesian Ridgeback pups will not always be the answer. But it seems likely that the appropriate cash crops for asteroid colonies will often be products of specialized plant and animal breeding. Every asteroid colony must begin with a plant and animal breeding program aimed at the establishment of an ecology adapted to the local conditions. As a result, most colonies will possess varieties of plant and animal which are rare or nonexistent elsewhere. Just as on the North Pacific island, the isolation of an asteroid provides an ideal environment for maintaining purebred pedigrees. And every colony is likely to include not only a competent blacksmith but also an expert in genetic engineering.

In Chapter 9, I mentioned two new technologies of space propulsion which may flourish in the next century, laser propulsion and solar sails. A third new technology for propulsion is the mass-driver. My friend Gerard O'Neill invented the mass-driver and built a working model of it at his Space Studies Institute in Princeton. The mass-driver is a long magnetic accelerator which pushes little buckets down a straight track. You put into the buckets any material which happens to be cheap and available. The contents of the buckets are thrown out into space at high speed. Like the exhaust of a rocket, they exert a thrust on the vehicle to which the mass-driver is attached. Unlike a rocket, the mass-driver can keep on running forever provided it is supplied with electric power. The Sun will supply enough power to keep a mass-driver running anywhere within the inner solar system. Ships propelled by mass-drivers would be an efficient and economical means of transport for voyagers moving around in the asteroid archipelago.

Like the Polynesian canoe, the mass-driver ship will be slow but will cover long distances cheaply. After arriving at an asteroid, the voyagers can use the local soil to reload the ship with propellant mass for the next leg of their journey. The main disadvantage of using mass-drivers in this way is that the bucketloads of dirt will gradually fill the space around the asteroids with rings of asteroidal dust like an attenuated version of the rings of Saturn. But the problem of the pollution of the solar system with dust has a remedy if ever it becomes serious. Oxygen is a major component of all asteroidal soils, the other major components being metals and silicon. Every asteroid colony will set up a process, either chemical or biological, for separating oxygen from the soil for breathing. Every colony is likely to maintain a large reserve supply of oxygen for emergencies. And liquid oxygen would be an ideal non-polluting propellant for use in mass-drivers. The liquid oxygen thrown out into space would quickly evaporate and be carried harmlessly outward away from the Sun by the natural solar wind.

We cannot tell today whether mass-drivers, laser propulsion, solar sails or other propulsion systems still to be invented will prevail in the economic competition of the future. Probably each of these modes of propulsion will find its appropriate ecological niche. Each of these systems has the potential of being enormously cheaper than present-day chemical rockets. The space technology of today is absurdly expensive, for a variety of reasons. With today's technology there is not much incentive to reduce the costs of propulsion, since the payloads of most missions cost more than the propulsion systems. So long as space missions cost thousands of dollars per pound of payload, space colonies will remain an idle dream. The expansion of life into space will begin in earnest only when we are in command of radically cheaper technologies. Fortunately, the mass-driver and the laser and the solar sail give us a promise of radically cheaper propulsion, and genetic engineering and artificial intelligence give us a promise of radically cheaper payloads.

The next hundred years will be a period of transition between the metal-and-silicon technology of today and the enzyme-and-nerve technology of tomorrow. The enzyme-and-nerve technology will be the result of combining the tools of genetic engineering and artificial intelligence. We cannot hope to predict the concrete forms in which a mature enzyme-and-nerve technology would express itself. When I think of the space technology of tomorrow, I think of three concrete images in particular. First, the Martian potato, a succulent plant that lives deep underground, its roots penetrating layers of subterranean ice while its shoots gather carbon dioxide and sunlight on the surface under the protection of a self-generated greenhouse. Second, the comet creeper, a warm-blooded vine which spreads like a weed over the surfaces of comets and keeps itself warm with super-insulating fur as soft as sable. Third, the space butterfly which I mentioned in Chapter 9, a creature truly at home in space, acting as our agent in exploration and reconnaissance, carrying pollen and information from world to world just as terrestrial insects carry pollen from flower to flower. It is easy to dream of other inhabitants of the celestial zoo, other images of a universe coming to life. The Martian potato, the comet creeper and the space butterfly are merely symbols, intended, like the pictures in a medieval bestiary, to edify rather than to enlighten.

I will not try to describe in detail the ways which space technology might follow when we look more than a hundred years into the future. Beyond the space butterfly is a further evolution as complicated and as unpredictable as the evolution of life on Earth. Only one thing is certain. The evolution of life in the universe will be like the evolution of life on Earth, an unfolding of weird and improbable patterns, an unfolding of ever-increasing richness and diversity.

When life spreads out and diversifies in the universe, adapting itself to a spectrum of environments far wider than any one planet can encompass, the human species will one day find itself faced with the most momentous choice that we have had to make since the days when our ancestors came down from the trees in Africa and left their cousins the chimpanzees behind. We will have to choose, either to remain one species united by a common bodily shape as well as by a common history, or to let ourselves diversify as the other species of plants and animals will diversify. Shall we be forever one people, or shall we be a million intelligent species exploring diverse ways of living in a million different places across the galaxy? This is the great question which will soon be upon us. Fortunately, it is not the responsibility of this generation to answer it.

17

BUTTERFLIES AGAIN

To bring the dead to life
Is no great magic.
Few are wholly dead:
Blow on a dead man's embers
And a live flame will start.

Robert Graves's poem catches in a few words one of the central themes of human history. Our literature from the oldest epics to Marcel Proust, our religions from ancient Egypt to the Latter-Day Saints, our political arrangements from Solomon's Kingdom to the United States Constitution, all are concerned in one way or another with giving an extended life to the past. Our holy books tell us to give honor to our parents. Our scientific curiosity leads us to go wandering in time as well as in space. Homo sapiens is not only, as Ben Finney observed, the exploring animal, the species that invented geography. We invented history and archeology too. Our museums and libraries and art galleries are filled with holy relics of the past. The great Einstein one day walked by mistake into a schoolboys' changing room where sports clothes were hanging on pegs underneath tablets recording the names of past generations of boys. "Ach," said Einstein, "I understand. The spirit of the departed passes into the trousers of the living."

Our awareness of the past brings with it an awareness of our own mortality. The longer and the more far-reaching our vision of the past, the shorter and the more ephemeral becomes the span of individual human life. Robert Graves's poem describes also what happens to us when we are successful in bringing the dead to life:

> So grant him life, but reckon
> That the grave which housed him
> May not be empty now:
> You in his spotted garments
> Must yourself lie wrapped.

Our technology is giving us progressively greater power to keep alive our ancestors' ghosts. First the invention of writing allowed us to preserve their words. Painting and photography allowed us to preserve their faces. The phonograph preserves their voices and the videotape recorder preserves their movement and gestures. But this is only the beginning. Soon we shall acquire the technology to preserve a permanent record of the sequence of bases in the DNA of their cells. This means that we shall be able, if we wish, to carry the magic a stage further, to reconstruct from the DNA sequence a genetic copy or clone of the ancestor. After that, perhaps, will come the technology to read the memory traces that record the experiences of a lifetime in the ancestor's brain. And then, perhaps, the technology to play back the ancestor's memories and feelings into the consciousness of the living. At that point the distinction between living and dead, present and past, will become blurred. It will be hard to tell who is the ancestor and who is the descendant, who is the one blowing on the embers and who is the one lying wrapped in spotted garments in the grave.

Human beings are so constituted that we take for granted the fact that a direct awareness of our past selves is preserved through a lifetime of experience. We take for granted the durability of the individual self. We think of death as an interruption of the natural order of things, an extraneous event requiring explanation. We think of bringing the dead to life as an exercise in magic. But if we stand back from our customary presuppositions and look at the human situation objectively, we see that the preservation of memories within an individual human life from childhood to old age is as great an exercise in magic as the transfer of memories from the dead to the living. We know by introspection that this magic works, that we retain, through sixty years of buffeting in the turbulent stream of time, a direct awareness of our past. The huge dappled-gray rocking horse on which I rode as a three-year-old still seems to tower over my head as I contemplate it as a sixty-three-year-old. How the magic works, how the traces of the rocking horse were laid down and preserved and recalled after sixty years, is still a dark mystery. But we shall learn. We shall penetrate the mystery. And when we have understood the magic, it will be magic no longer. It will then be only a new science for us to explore, a new technology for us to exploit.

When I look to the future of humanity beyond the twenty-first century, I see on my list of things to come the extension of our inquisitiveness from the objective domain of science to the subjective domain of feeling and memory. Homo sapiens, the exploring animal, will not be content with merely physical exploration. Our curiosity will drive us to explore the dimensions of mind as vigorously as we explore the dimensions of space and time. For every pioneer who explores a new asteroid or a new planet, there will be another pioneer who explores from the inside the minds of our fellow passengers on planet Earth. It is our nature to strive to explore everything, alive and dead, present and past and future. When once the technology exists to read and write memories from one mind into another, the age of mental exploration will begin in earnest. Instead of admiring the beauties of nature from the outside, we will look at nature directly through the eyes of the elephant, the eagle and the whale. We will be able, through the magic of science, to feel in our own minds the pride of the peacock and the wrath of the lion. That magic is no greater than the magic that enables me to see the rocking horse through the eyes of the child who rode it sixty years ago.

Every speculation concerning the long-range future of humanity must end, as mine is ending, in fantasy. Nobody can hope to foresee the quirks of history, the peculiar accidents of human wisdom and folly, that will determine our real future. As Desmond Bernal wrote in 1929, "There are two futures, the future of desire and the future of fate, and man's reason has never learned to separate them." The future which I have been sketching here is the future of desire. The future of fate is harder to discern. Nobody has imagined the future of fate with greater artistry than H. G. Wells in his fantasy *The Time Machine,* published in 1895. Wells imagined the human species split in two, the spark of reason dulled and the sense of purpose extinguished. His two species, the degenerate descendants of the upper and lower classes of Victorian England, are caught in an evolutionary dead end without hope of escape. The lower class, living underground like rats, has retained enough manual dexterity to keep the machinery running. The upper class, living aimlessly on the surface, is maintained like cattle in happy ignorance until the night comes when it is taken down to be butchered and eaten by the tunnel-dwellers. This nightmare is indeed a possible future for mankind. Wells held it up as a mirror in which his contemporaries could see reflected the ugliness and the injustice of their own society. It still may serve us as a mirror to reflect the failures of our society today.

But Wells was not content to be only a prophet of despair. He was also a prophet of hope. He made it his business to hold up a variety of mirrors, visions of hope as well as visions of despair, futures of desire as well as futures of fate. Seven years after publishing *The Time Machine,* he gave a lecture at the Royal Institution in London with the title "The Discovery of the Future." This was his first attempt to portray a future of desire. Here is a little piece of his lecture. He is describing a vision which the triumphs and calamities of the intervening years from 1902 to 1987 have in no way invalidated:

> We look back through countless millions of years and see the great will to live struggling out of the intertidal slime, struggling from shape to shape and from power to power, crawling and then walking confidently upon the land, struggling generation after generation to master the air, creeping down into the darkness of the deep; we see it turn upon itself in rage and hunger and reshape itself anew, we watch it draw nearer and more akin to us, expanding, elaborating itself, pursuing its relentless inconceivable purpose, until at last it reaches us and its being beats through our brains and arteries, throbs and thunders in our battleships, roars through our cities, sings in our music and flowers in our art. And when, from that retrospect, we turn again towards the future, surely any thought of finality, any millennial settlement of cultured persons, has vanished from our minds. The fact that man is not final is the great unmanageable disturbing fact that rises upon us in the scientific discovery of the future, and to my mind at any rate the question what is to come *after* man is the most persistently fascinating and the most insoluble question in the whole world. Of course we have no answer. Such imaginations as we have refuse to rise to the task.

Wells was well aware of the precariousness of human life, both for the individual and for the species.

> It is conceivable that some great unexpected mass of matter should presently rush upon us out of space, whirl sun and planets aside like dead leaves before the breeze, and collide with and utterly destroy every spark of life upon this earth. There is nothing in science to show why such a thing should not be. It is conceivable, too, that some pestilence may presently appear, some new disease, that will destroy, not 10 or 15 or 20 per cent of the earth's inhabitants as pestilences have done in the past, but 100 per cent, and so end our race. No one, speaking from

scientific grounds alone, can say, that cannot be. There may arise new animals to prey upon us by land and sea, and there may come some drug or a wrecking madness into the minds of men. And finally there is the reasonable certainty that this sun of ours must some day radiate itself toward extinction; that at least must happen, until some day this earth of ours, tideless and slow moving, will be dead and frozen, and all that has lived upon it will be frozen out and done with. There surely man must end. That of all such nightmares is the most insistently convincing. And yet one doesn't believe it. At least I do not. And I do not believe in these things because I have come to believe in certain other things—in the coherency and purpose in the world and in the greatness of human destiny. Worlds may freeze and suns may perish, but there stirs something within us now that can never die again.

We are in the beginning of the greatest change that humanity has ever undergone. There is no shock, no epoch-making incident, but then there is no shock at a cloudy daybreak. At no point can we say, here it commences, now, last minute was night and this is morning. But insensibly we are in the day. What we can see and imagine gives us a measure and gives us faith for what surpasses the imagination.

Wells ends his discourse with the little word "faith." His visions of the future were mostly pessimistic, from the rottenness of *Tonobungay* to the final decadence of *The Time Machine*. But his soul rebelled against the pessimistic visions of his mind. And so he found it necessary to supplement his science with a little bit of faith. As the mathematician Blaise Pascal wrote 240 years earlier: "Le Coeur a ses raisons que la raison ne connaît point." All of us who think and care about the human situation must be impelled by some kind of faith as well as by scientific knowledge. It is good to follow the example of Wells and make the tenets of our faith as conscious and as explicit as possible. The concluding section of this last chapter summarizes my thoughts about the proper limits of faith and reason.

My own faith, as I described it in Chapter 6, is similar to the faith of Wells. I believe that we are here to some purpose, that the purpose has something to do with the future, and that it transcends altogether the limits of our present knowledge and understanding. I do not wish to go beyond this simple statement into a discussion of theology. My ignorance of theology would quickly become obvious. If you like, you can call the transcendent purpose God. If it is God, it is a Socinian God, inherent in the universe and growing in power and knowledge as the universe unfolds. Our minds are not only expressions of its purpose but are also contributions to its growth.

In the no-man's-land between science and theology, there are five specific points at which faith and reason may appear to clash. The five points are the origin of life, the human experience of free will, the prohibition of teleological explanations in science, the argument from design as an explanatory principle, and the question of ultimate aims. Each of these points could be the subject of a whole chapter, but fortunately I have only a few pages for all five. I will deal with each of them in turn as well as I can in a few lines.

First, the origin of life. This is not the most difficult problem from a philosophical point of view. Life in its earliest stages was little removed from ordinary chemistry. We can at least imagine life originating by ordinary processes which chemists know how to calculate. Much more serious problems for philosophy arise at a later stage with the development of mind and consciousness and language. As the physicist Wigner once said: "Where in the Schrödinger equation do you put the joy of being alive?" The problem with the origin of life is only this: How do you reconcile a theory which makes life originate by a process of chance with the doctrine that life is a part of God's plan for the universe? There are three possible answers to this question. Answer 1. Deny that God has a plan and say that everything is accidental. This is the answer of Jacques Monod, and of the majority of modern biologists. But then Wigner will ask: Is consciousness also an accident? Answer 2. Deny that chance exists and say that God knows how the dice will fall. This is the answer of Einstein, who believed that chance is a human concept arising from our ignorance of the exact working of nature. But then, why do statistical laws play such a fundamental role in physics, if chance is only a cover for our ignorance? Answer 3. Say that chance exists because God shares our ignorance. This is the answer of Hartshorne, the Socinian heresy. God is not omniscient. He grows with the universe and learns as it develops. Chance is a part of his plan. He uses it as we do to achieve his ends.

The second clash between faith and reason is the problem of free will. It was formulated most clearly by Schrödinger in the epilogue at the end of his little book *What Is Life?* The problem is to reconcile the direct human experience of free will with a belief in scientific causality. Here again we have the same three alternative answers to deal with the conflict. But now both narrow-minded science and narrow-minded theology stand opposed to free will. The Jacques Monod view of the universe as pure "Chance and Necessity" denies free will. The orthodox theology of an omniscient and omnipotent God also denies it. For those of us who would like to believe both in God and in free will, the Socinian answer is the best way out. The philosophical problems of chance and of free will are closely related. The Socinian theology deals with both together. Free will is the coupling of a human mind to otherwise random processes inside a brain. God's will is the coupling of a universal mind to otherwise random processes in the world at large.

My third problem is the problem of forbidden teleology, the conflict between human notions of purpose and the operational rules of science. Science does not accept Aristotelian styles of explanation, that stone falls because its nature is earthy and so it likes to be on Earth, or that man's brain evolves because man's nature is to be intelligent. Within science, all causes must be local and instrumental. Purpose is not acceptable as an explanation of scientific phenomena. Action at a distance, either in space or time, is forbidden. Especially, teleological influences of final goals upon phenomena are forbidden. How do we reconcile this prohibition with our human experience of purpose and with our faith in a universal purpose? I make the reconciliation possible by restricting the scope of science. The choice of laws of nature, and the choice of initial conditions for the universe, are questions belonging to meta-science and not to science. Science is restricted to the explanation of phenomena within the universe. Teleology is not forbidden when explanations go beyond science into meta-science.

The most familiar example of a meta-scientific explanation is the so-called Anthropic Principle. The Anthropic Principle says that laws of nature are explained if it can be established that they must be as they are in order to allow the existence of theoretical physicists to speculate about them. We know that theoretical physicists exist: ergo, the laws of nature must be such as to allow their existence. This mode of explanation is frankly teleological. It leads to non-trivial consequences, restrictions on the possible building blocks of the universe, which I have no space to discuss in detail. Many scientists dislike the Anthropic Principle because it seems to be a throwback to a pre-Copernican, Aristotelian style of reasoning. It seems to imply an anthropocentric view of the cosmos. Whether you like the Anthropic Principle or not is a matter of taste. I personally find it illuminating. It accords with the spirit of modern science that we have two complementary styles of explanation, the teleological style allowing a role for purpose in the universe at large, and the non-teleological style excluding purpose from phenomena within the strict jurisdiction of science.

The argument from design is the fourth on my short list of philosophical problems. The argument was one of the classic proofs of the existence of God. The existence of a watch implies the existence of a watchmaker. This argument was at the heart of the battle between creationists and evolutionists in nineteenth-century biology. The evolutionists won the battle. Random genetic variations plus Darwinian selection were shown to be sufficient causes of biological evolution. The argument from design was excluded from science because it makes use of teleological causes. For a hundred years the biologists have been zealously stamping out all attempts to revive the old creationist doctrines. Nevertheless, the argument from design still has some merit as a philosophical principle. I propose that we allow the argument from design the same status as the Anthropic Principle, expelled from science but tolerated in meta-science.

The argument from design is a theological and not a scientific argument. It is a mistake to try to squeeze theology into the mold of science. I consider the argument from design to be valid in the following sense. The universe shows evidence of the operations of mind on three levels. The first level is the level of elementary physical processes in quantum mechanics. Matter in quantum mechanics is not an inert substance but an active agent, constantly making choices between alternative possibilities according to probabilistic laws. Every quantum experiment forces nature to make choices. It appears that mind, as manifested by the capacity to make choices, is to some extent inherent in every electron. The second level at which we detect the operations of mind is the level of direct human experience. Our brains appear to be devices for the amplification of the mental component of the quantum choices made by molecules inside our heads. We are the second big step in the development of mind. Now comes the argument from design. There is evidence from peculiar features of the laws of nature that the universe as a whole is hospitable to the growth of mind. The argument here is merely an extension of the Anthropic Principle up to a universal scale. Therefore it is reasonable to believe in the existence of a third level of mind, a mental component of the universe. If we believe in this mental component and call it God, then we can say that we are small pieces of God's mental apparatus.

The last of the five philosophical problems is the problem of final aims. The problem here is to try to formulate some statement of the ultimate purpose of the universe. In other words, the problem is to read God's mind. Previous attempts to read God's mind have not been notably successful. One of the more penetrating of such attempts is recorded in the Book of Job. God's answer to Job out of the whirlwind was not encouraging. Nevertheless I stand in good company when I ask again the questions Job asked. Why do we suffer? Why is the world so unjust? What is the purpose of pain and tragedy? I would like to have answers to these questions, answers which are valid at our childish level of understanding even if they do not penetrate far into the mind of God. My answers are based on a hypothesis which is an extension both of the Anthropic Principle and of the argument from design. The hypothesis is that the universe is constructed according to a principle of maximum diversity. The principle of maximum diversity operates both at the physical and at the mental level. It says that the laws of nature and the initial conditions are such as to make the universe as interesting as possible. As a result, life is possible but not too easy. Always when things are dull, something new turns up to challenge us and to stop us from settling into a rut. Examples of things which make life difficult are all around us: comet impacts, ice ages, weapons, plagues, nuclear fission, computers, sex, sin and death. Not all challenges can be overcome, and so we have tragedy. Maximum diversity often leads to maximum stress. In the end we survive, but only by the skin of our teeth.

The expansion of life and of mankind into the universe will lead to a vast diversification of ecologies and of cultures. As in the past, so in the future, the extension of our living space will bring opportunities for tragedy as well as achievement. To this process of growth and diversification I see no end. It is useless for us to try to imagine the varieties of experience, physical and intellectual and religious, to which mankind may attain. To describe the metamorphosis of mankind as we embark on our immense journey into the universe, I return to the humble image of the butterfly. All that can be said was said long ago by Dante in Canto 10 of the *Purgatorio*:

O you proud Christians, wretched souls and small,
Who by the dim lights of your twisted minds
Believe you prosper even as you fall,
Can you not see that we are worms, each one
Born to become the angelic butterfly
That flies defenseless to the Judgment Throne?

BIBLIOGRAPHICAL NOTES

I have not attempted to make these notes complete or up to date. They contain only references which I found useful in preparing the lectures.

PREFACE

F. J. Dyson, "Origins of Life" (Cambridge: Cambridge University Press, 1985).

PART I EPIGRAPH

The quote from Lord Gifford's will was attached to my letter of appointment as Gifford Lecturer.

1. IN PRAISE OF DIVERSITY

This chapter is mostly taken from a talk on "Science and Religion," given in Detroit (September 1986) at a meeting organized by the Committee on Human Values of the National Conference of Catholic Bishops. See David M. Byers, ed. "Religion, Science and the Search for Wisdom," (Washington, D.C., United States Catholic Conference, 1987), pp. 47–62.

William James, "The Varieties of Religious Experience, A Study in Human Nature" (London: Longmans Green, 1937), Gifford Lectures given at Edinburgh, 1901–02.

H. G. Wells, "The Outline of History, being a Plain History of Life and Mankind" (London: George Newnes, 1920), 2 vols. My quotes are from Vol. 1, pp. 11 and 364.

For the Pastoral Letter, see note on chapter 11.

The "famous biologist" is Edward O. Wilson, who spoke on "Evolutionary Biology and Religion" at the Catholic bishops' meeting. See David M. Byers, ed., op. cit. pp. 82–90.

2. BUTTERFLIES AND SUPERSTRINGS

Michael B. Green, "Superstrings," *Scientific American,* 255: 48–60 (September 1986), is an account written for non-experts, but it assumes some knowledge of mathematics. Green is one of the two inventors of superstrings.

Stephen W. Hawking, "The Quantum Mechanics of Black Holes," *Scientific American,* 236: 34–40 (January 1977), explains the theory as simply as it can be explained without violating Bohr's rule. Bohr's rule says that one should not write more clearly than one can think.

Jan H. Oort, *Bulletin of the Astronomical Institutes of the Netherlands,* 11: 91–110 (1950), first described the Oort Cloud.

Walter Alvarez, *Eos,* 67: 649–658 (September 1986), reviews the evidence for comet showers up to spring 1986, with references to earlier papers. For a more recent summary see P. Hut et al., *Nature, 329,* 118–126 (September 1987).

Paul R. Weissman, *Sky and Telescope,* 73: 238–241 (March 1987), describes the history of the Oort Cloud and of the idea of comet showers.

H. N. Russell, R. S. Dugan, and J. Q. Stewart, "Astronomy, Volume I, The Solar System" (Boston: Ginn and Co., rev. ed. 1945), discusses the three planets: Vulcan, p. 358, Neptune, pp. 399–400, and Pluto, pp. 404–407.

3. MANCHESTER AND ATHENS

This chapter is a revised version of a talk given at the Sixteenth Nobel Conference at Gustavus Adolphus College, St. Peter, Minnesota, in 1980. The talk was published in the conference Proceedings, *The Aesthetic Dimension of Science,* ed. Deane W. Curtin (New York: Philosophical Library, 1982), pp. 41–62.

Emil Wiechert, "Die Theorie der Elektrodynamik und die Röntgensche Entdeckung," *Schriften der Physikalisch-Ökonomischen Gesellschaft zu Königsberg in Preussen,* 37: 1–48 (1896).

Benjamin Disraeli, *Coningsby, or the New Generation* (London: Longmans Green, 1844). The quotation is from p. 152 of the 1849 edition.

Arnold Thackray, "Natural Knowledge in Cultural Context: The Manchester Model," *American Historical Review,* 79: 672–709 (1974).

J. B. Birks, ed., *Rutherford at Manchester* (London: Heywood and Co., 1962).

S. Chandrasekhar, "Einstein and General Relativity: Historical Perspectives," *American Journal of Physics,* 47: 212–217 (1979).

Francis Crick, "On Protein Synthesis," *Symposium of the Society for Experimental Biology,* 12: 138–163 (1957).

John Wheeler, *Frontiers of Time* (Austin, Tex.: Center for Theoretical Physics, University of Texas, 1978), p. 13.

For the Huygens quotation, see Marjorie H. Nicolson, *Voyages to the Moon* (New York: Macmillan, 1948), pp. 60–62.

Frank Manuel, *The Religion of Isaac Newton* (London: Oxford University Press, 1974), pp. 99–102.

Michael Polanyi, "Personal Knowledge" (Chicago: University of Chicago Press, 1958), p. 404. This is the published version of Polanyi's Gifford Lectures given at Aberdeen in 1951–52.

4. HOW DID LIFE BEGIN?

This and the following chapter are an abbreviated version of my 1985 Tarner Lectures; see note on Preface.

James D. Watson, *The Molecular Biology of the Gene* (New York: Benjamin, 1965), is a standard textbook of molecular biology.

Frank Close, *The Cosmic Onion: Quarks and the Nature of the Universe* (New York: American Institute of Physics, 1986), is a good introduction to particle physics.

Leslie E. Orgel, "RNA Catalysis and the Origins of Life," *Journal of Theoretical Biology,* 123: 127–149 (1986), is a review of the new experimental evidence for hardware functions performed by nucleic acids. Orgel argues that the evidence supports a primary role for RNA in the origin of life. But I am unconvinced. We have known for a long time that RNA has an even more essential hardware function, the binding and transport of amino acids by transfer RNA. The hardware functions of RNA in modern cells are consistent with all theories of the origin of life. For an even more recent contribution to this debate see G. F. Joyce et al., *Proc. Nat. Acad. Sci. 84,* 4398–4402 (July 1987).

Erwin Schrödinger, *What Is Life? The Physical Aspect of the Living Cell* (Cambridge: Cambridge University Press, 1944).

John von Neumann, *The General and Logical Theory of Automata,* in J. von Neumann, *Collected Works,* ed. A. H. Taub (New York: Macmillan, 1961–63), Vol. 5, pp. 288–328.

M. Eigen, W. Gardiner, P. Schuster, and R. Winckler-Oswatitch, "The Origin of Genetic Information," *Scientific American,* 244: 88–118 (April 1981).

L. Margulis, *Symbiosis in Cell Evolution* (San Francisco: Freeman and Co., 1981).

M. Kimura, *The Neutral Theory of Molecular Evolution* (New York: Cambridge University Press, 1983).

5. WHY IS LIFE SO COMPLICATED?

A. I. Oparin, *The Origin of Life on the Earth,* 3rd ed., translated by Ann Synge (Edinburgh: Oliver & Boyd, 1957).

A. G. Cairns-Smith, *Genetic Takeover and the Mineral Origins of Life* (New York: Cambridge University Press, 1982).

F. J. Dyson, "A Model for the Origin of Life," *Journal of Molecular Evolution,* 18: 344–350 (1982); technical description of my model of the Oparin theory.

R. Dawkins, *The Selfish Gene* (New York: Oxford University Press, 1976).

E. O. Wilson, remarks quoted by R. Lewin in *Science,* 216: 1091–1092 (1982).

6. HOW WILL IT ALL END?

This chapter is based on a set of James Arthur Lectures on Time and Its Mysteries, given at New York University in 1978. The lectures were published with the title "Time Without End: Physics and Biology in an Open Universe," in *Reviews of Modern Physics,* 51: 447–460 (1979). Jamal N. Islam, *The Ultimate Fate of the Universe* (Cambridge: Cambridge University Press, 1983), gives a more complete and up-to-date account of the subject.

Steven Weinberg, *The First Three Minutes* (New York: Basic Books, 1977), pp. 131–132, 154.

Jacques Monod, *Chance and Necessity,* translated by A. Wainhouse (New York: Knopf, 1971), p. 176.

Konstantin Tsiolkovsky, *Dreams of Earth and Sky*, ed. B. N. Vorobyeva (Moscow: USSR Academy of Sciences, 1959), pp. 40–41 (my translation).

A. E. Housman, *More Poems* (New York: Knopf, 1936), p. 64.

Ivor Gurney's poem appears in Paul Fussell, *The Boy Scout Handbook and Other Observations* (New York: Oxford University Press, 1982), pp. 173–174. Also in P. J. Kavanaugh, ed. *Collected Poems of Ivor Gurney* (Oxford, Oxford University Press, 1982).

J. D. Bernal, *The World, the Flesh and the Devil: An Enquiry into the Future of the Three Enemies of the Rational Soul,* 2nd ed. (Bloomington, Ind.: Indiana University Press, 1969), pp. 46, 63.

J. B. S. Haldane, *Daedalus, or Science and the Future* (London: Kegan Paul, 1924), p. 92.

Thomas Wright, "An Original Theory or New Hypothesis of the Universe," facsimile reprint (New York: American Elsevier, 1971) pp. 76, 83–84.

Charles Hartshorne, *The Divine Relativity: A Social Conception of God* (New Haven: Yale University Press, 1948). After examining Hartshorne's book, I realized that the views which he modestly labeled "Socinian" should more properly be termed "Hartshornian."

Emlyn Williams, *Spring 1600, a Comedy in Three Acts* (London: Heinemann, 1946).

PART 2   EPIGRAPH

D. H. Lawrence, "Study of Thomas Hardy," in *Phoenix: The Posthumous Papers of D. H. Lawrence,* ed. E. D. McDonald (New York: Viking Press, 1936), pp. 425–426.

7. ROOTS

This chapter is based on an Oration at the Phi Beta Kappa Literary Exercises, Harvard University, June 1986, published in the *Harvard Magazine,* 88: 17–20 (July 1986).

Alex Haley, *Roots* (Garden City, N.Y.: Doubleday, 1976).

For Thomas Oliver, see *Dictionary of National Biography* (London: Oxford University Press, 1917–), Vol. 14, pp. 1045–1046. For

Peter Oliver, *Dictionary of American Biography* (New York, Scribner's, 1934), Vol. 14, pp. 22–23.

Richard Hakluyt, *The Principal Navigations, Voyages, Traffiques and Discoveries of the English Nation,* 2nd ed., 3 vols., 1598–1600; reprinted in 12 vols. (Glasgow: James MacLehose & Sons, 1903), Vol. 1, pp. 34–35, 66.

Geoffrey Keynes, ed., *Poetry and Prose of William Blake* (London: Nonesuch Press, 1939), pp. 90, 201, 202, 206, 761.

8. QUICK IS BEAUTIFUL

This chapter is based on a Tykociner Memorial Lecture given at the University of Illinois in 1981, published by the University of Illinois as a booklet with the title *Quick Is Beautiful.*

Lynn White, Jr., "Technology Assessment from the Stance of a Medieval Historian," *American Historical Review,* 79: 1–13 (1974).

Michael I. Pupin, *From Immigrant to Inventor* (New York: Scribner's, 1960), Foreword, pp. 9–10.

For a survey of the HTGR and other gas-cooled reactors, see G. Melese-d'Hopital and M. Simnad, "Status of Helium-cooled Nuclear Reactors," *Energy,* 2: 211–239 (1977).

For ice ponds, see Theodore B. Taylor, "Ice-Pond Technology" (Unpublished report, September 1985).

9. SCIENCE AND SPACE

This chapter is based on a lecture given at the National Academy of Sciences in Washington, D.C., October 1982, published in Allan A. Needell, ed., *The First 25 Years in Space* (Washington, D.C.: Smithsonian Institution Press, 1983), pp. 90–106. Another version is published in S. L. Shapiro and S. A. Teukolsky, eds., *Highlights of Modern Astrophysics, Concepts and Controversies* (New York: Wiley, 1986), pp. 367–382. Parts of it also appeared in *Science 85* (November 1985), 127–130.

Lyman Spitzer, *Searching Between the Stars* (New Haven: Yale University Press, 1982), describes the Copernicus mission (pp. 57–68) and the wealth of information which it collected about the interstellar gas (pp. 88–95, 98–107, 121–124). Spitzer's account emphasizes the complementarity between the Copernicus ultraviolet observations, the ground-based millimeter-wave observations, and the space-based X-ray observations. All three types of observation were essential in constructing the modern picture of the interstellar gas with its diverse regions, some cold, some hot, and some super-hot.

For details of the Crab Nebula occultation observation see S. Bowyer, E. Byram, T. Chubb and H. Friedman, *Science,* 146, 912–916 (1964).

For details of the Hipparcos mission, see M. A. C. Perryman, *Ad Astra Hipparcos: The European Space Agency's Astrometry Mission,* ESA Report BR-24 (June 1985).

For laser propulsion, see F. V. Bunkin and A. M. Prokhorov, "Use of a Laser Energy Source in Producing a Reactive Thrust," *Soviet Physics Uspekhi,* 19: 561–573 (1976), a review of Soviet and Western work with references to the earlier literature.

For solar sails and other unorthodox space vehicles, see "Discussion Meeting on Gossamer Spacecraft (Ultralightweight Spacecraft): Final Report," ed. R. G. Brereton (NASA Jet Propulsion Laboratory Publication, 80–26, May 1980).

10. ENGINEERS' DREAMS

This chapter is based on a lecture given at Analog Devices, Norwood, Massachusetts, in March 1986. Chapter 9 was written before the Challenger accident, chapter 10 after the accident. The argument of chapter 9 remains as valid after the accident as it was before.

Willy Ley, *Engineers' Dreams* (New York: Viking Press, 1954).

Apsley Cherry-Garrard, *The Worst Journey in the World: Antarctic 1910–1913* (London: Penguin Books, 1937), pp. 534–535.

For STIG, see Robert H. Williams and Eric D. Larson, "Steam-injected Gas Turbines and Electric Utility Planning," *IEEE Technology and Society Magazine,* 29–38 (March 1986).

11. THE BALANCE OF POWER

I have rearranged four Gifford Lectures (#s 7, 8, 9 and 12) into the seven chapters 11–17.

The Einstein quotations come from Otto Nathan and Heinz Norden, eds, *Einstein on Peace* (New York: Simon and Schuster, 1960), pp. 190, 448, 476, and from Jamie Sayen, *Einstein in America* (New York: Crown Publishers, 1985), p. 239.

Alexander Hamilton, James Madison, and John Jay, "The Federalist," *Great Books of the Western World,* Vol. 43 (Founders' Edition, Chicago: Encyclopaedia Britannica, 1952), pp. 37, 39–41.

Pastoral Letter of the United States Bishops, "The Challenge of Peace: God's Promise and Our Response," *Origins,* 13: 1–32 (Washington, D.C., National Catholic News Service, 1983).

The statement that the tribe of Israel lived without an army and without war for two thousand years may be incorrect. See Arthur Koestler, *The Thirteenth Tribe: The Khazar Empire and Its Heritage* (London: Pan Books, 1976), for a contrary view.

12. STAR WARS

In April 1985 I gave the Albert Pick Lecture at the University of Chicago with the title "Star Wars and Austrianization and Nuclear Winter." The Pick Lecture grew into Gifford Lectures 7, 8, 9, and into chapters 12, 13, 15, 16 of this book.

Sidney Drell, Philip Farley, and David Holloway, *The Reagan Strategic Defense Initiative: A Technical, Political and Arms Control Assessment* (Stanford University, Ca.: Center for International Security and Arms Control, July 1984). The text of President Reagan's announcement is Appendix A, pp. 101–103.

Harvey Lynch, *Technical Evaluation of Offensive Uses of SDI* (Stanford University, Ca.: Center for International Security and Arms Control, February 1987), analyzes the use of an SDI laser in orbit to "strike the Soviet Politburo assembled on the reviewing stand during a parade through Red Square" (pp. 39–40). Lynch assumes that lasers in orbit "function as advertised." He does not discuss offensive uses of more realistic versions of SDI.

Stefan Forss, "The Soviet Offensive on the Finnish Front in the Summer of 1944: Implications for the No-First-Use Debate" (Unpublished, 1985). I am grateful to Professor Forss for this article and for several letters on the same subject. Also for the text of a lecture, "Finland as a European Country," given by President Kekkonen at the Übersee Club in Hamburg in May 1979, making explicit the relevance of Finnish experience to the situation of Germany.

For a thorough and detailed technical assessment of Star Wars technology, see the "Report to the American Physical Society of the Study Group on Science and Technology of Directed Energy

Weapons," *Reviews of Modern Physics,* 59, No. 3, Part II (July 1987), pp. S1–S202.

13. THE EXAMPLE OF AUSTRIA

Gerald Stourzh, *Geschichte des Staatsvertrages 1945–1955: Österreichs Weg zur Neutralität* (Graz: Verlag Styria, 2nd enlarged ed. 1980). The text of the State Treaty in its initial (1947) and final (1955) versions occupies pp. 243–316. The text of the Constitutional Law in English and German is on pp. 239–240. For the history of the decisive negotiation in Moscow in April 1955, see pp. 142–163.

For a briefer account in English, see Alfred Verdross, *The Permanent Neutrality of Austria,* translated by C. Kessler, (Vienna: Verlag für Geschichte und Politik, 1978).

14. CAMELS AND SWORDS

A small part of this chapter was published as "First Word" in *Omni* magazine (July 1984).

Pastoral Letter of the United States Bishops, *op. cit.,* pp. 17, 30.

Richard W. Bulliet, *The Camel and the Wheel* (Cambridge, Mass.: Harvard University Press, 1975).

Noel Perrin, *Giving Up the Gun: Japan's Reversion to the Sword, 1543–1879* (Boulder, Col.: Shambala Publications, 1980).

Minoru Oda, "Old Soldier's Monologue" (Unpublished manuscript), an informal history of X-ray satellite missions at the Institute of Space and Astronautical Science.

The Stalin quotation is taken from George Kennan, *The Nuclear Delusion: Soviet-American Relations in the Atomic Age* (New York: Pantheon Books, 1982), p. 71.

15. NUCLEAR WINTER

Part of this chapter was published in the *New York Times Magazine,* April 5, 1987, with the title "Demystifying the Bomb."

Paul R. Ehrlich, Carl Sagan, Donald Kennedy, and Walter O. Roberts, *The Cold and the Dark: The World after Nuclear War* (New York: Norton, 1984), a collection of papers from the first Nuclear Winter conferences in 1983. It sets the agenda for later discussions of nuclear winter. An extensive literature has grown up in response to it. As examples of the official response, see *The Effects on the Atmosphere of a Major Nuclear Exchange* (Washington, D.C.: National Academy Press, 1985), written by a United States National Academy of Sciences committee, or *Environmental Consequences of Nuclear War*, SCOPE, 28, Vol. 1. *Physical and Atmospheric Effects* (New York: Wiley, 1985), written by a committee of the International Council of Scientific Unions. The effects of wet soot on radiation transport are given only a brief mention in these reports (pp. 104–105 of the Academy report, p. 117 of the SCOPE report).

Robinson Jeffers, *The Double Axe and Other Poems, Including Eleven Suppressed Poems* (New York: Liveright, 1977), p. 164. The poem "Curb Science?" is one of the eleven poems suppressed by Random House editors when *The Double Axe* was published in 1948.

Marc Kaminsky, *The Road from Hiroshima* (New York: Simon and Schuster, 1984), pp. 107–108.

16. THE TWENTY-FIRST CENTURY

Two passages from this chapter were published in *Omni* magazine (February 1986 and June 1986).

J. B. S. Haldane, *Daedalus, or Science and the Future, op. cit.*, pp. 59–63, 89–90.

Sir James Lighthill, *Artificial Intelligence, a General Survey* (London: Science Research Council Report, 1972), pp. 27, 38.

B. R. Finney and E. M. Jones, eds., *Interstellar Migration and the Human Experience* (Berkeley: University of California Press, 1985). More directly relevant to this chapter is an earlier article by Finney and Jones, "From Africa to the Stars: The Evolution of the Exploring Animal" (Los Alamos National Laboratory, Report LA-UR-83-929, 1983).

For the mass-driver, see G. K. O'Neill, "The Colonization of Space," *Physics Today*, 27: 32–40 (1974). Reports on the performance of working models are in the quarterly newsletter of the Space Studies Institute, "SSI Update" (P.O. Box 82, Princeton, N.J., 08540). My version of the twenty-first century owes a great deal to G. K. O'Neill, *2081, A Hopeful View of the Human Future* (New York: Simon and Schuster, 1981).

17. BUTTERFLIES AGAIN

Robert Graves, "To Bring the Dead to Life," in *Faber Book of Modern Verse* (London: Faber and Faber, 1936), p. 233. This was the first publication of the poem (see p. 347).

For the Einstein story, see John G. Griffith "Albert Einstein at Winchester 1931" in *The Trusty Servant* (Winchester, Engl.: Culverlands Press, 1986), no. 62, p. 5. The exact words of Einstein were: "Ach! Ich verstehe. Der Geist der Gestorbenen geht in die Beinkleider der Lebenden hinüber." The translation is by F. A. Lindemann, later Lord Cherwell, who was acting as interpreter for the benefit of Griffith.

J. D. Bernal, *The World, the Flesh and the Devil, op. cit.,* p. 3.

H. G. Wells, "The Discovery of the Future," *Nature*, 65: 326–331 (1902).

The two questions of Wigner are quoted from private conversations. For Wigner's views on the mind-body problem, see Eugene P. Wigner, *Symmetries and Reflections* (Bloomington, Ind.: Indiana University Press, 1967), chapters 13–14, pp. 171–199.

E. Schrödinger, *What Is Life?, op. cit.,* pp. 87–91.

For the anthropic principle, see J. D. Barrow and F. J. Tipler, *The Anthropic Cosmological Principle* (New York: Oxford University Press, 1986).

Dante Alighieri, *The Divine Comedy,* translated by John Ciardi (New York: Norton, 1977), p. 241.

INDEX

COPYRIGHT
ACKNOWLEDGMENTS

Praise for *Infinite in All Directions*

"Freeman J. Dyson is an original, versatile, and thoughtful man, and what he writes is always interesting. But he is no cold scientist. He has a love of literature and history and a distinct talent for writing. Sometimes his writing has its elements of poetry. Perhaps Mr. Dyson's display of ideas has, indeed, elements in common with his beloved butterfly—beautiful, intricate, and something to be admired, but flitting delicately here and there. This is a book to be read, savored, and appreciated."　　—*New York Times*

"The bedazzled reader emerges feeing like he's been in a metaphysical Maytag on spin cycle—his perspective on man, God, and the cosmos permanently altered. Dyson [is] neuron for neuron, one of the most formidably provocative minds in American life. Throughout, Dyson's language, reminiscent of Orwell's, is eloquently plain, wrought with the unaffected grace of a man certain that he has something important to say. Exuberantly stimulating."
—*Washington Post Book World*

"Dyson makes a spirited stand. To observe a mind uncommonly endowed with dexterity and knowledge hop from subject to subject is exhilarating. Dyson inspires the same awe he reports at watching a butterfly emerge from its chrysalis and fly away."　　　　　　—*Time*

"Thoughtful, substantial, and readable."
—*Christian Science Monitor*

"[One of] the world's great theoretical physicists, Dyson permits himself to dip into all sorts of matters, scientific and not. He explains, in a way that is understandable even to someone who has spent his life being crunched by numbers rather than crunching them, what past and recent scientific theories tell us about the beginning, ending, and present state of the universe."   —USA Today

"A great book. . . . Dyson has an agile, ingenious mind. Highly recommended to all sorts and conditions of readers, who will be delighted, instructed, and set to pondering."
                                                              —Choice